SpringerBriefs in Modern Perspectives on Disability Research

This book series on disability research is a comprehensive collection of research on disability and related issues. The series is designed to promote interdisciplinary collaboration and exchange, bringing together scholars and practitioners from different fields to share their perspectives and insights. Disability research is an interdisciplinary field that examines the social, cultural, historical, and political dimensions of disability. It encompasses a wide range of topics, including disability rights, accessibility, assistive technologies, healthcare, education, employment, and social welfare. Disability research scholars employ a range of theoretical and methodological approaches to understand the experiences of people with disabilities, as well as the ways in which disability intersects with other social identities such as race, gender, sexuality, and class.

The series seeks to advance knowledge and understanding of disability by publishing rigorous, innovative, and relevant research. It aims to promote disability rights and social justice by highlighting the ways in which people with disabilities are marginalized and discriminated against in society, and advocating for greater social inclusion and accessibility. The series also seeks to inform policy and practice by disseminating research findings that can help to shape policy decisions and contribute to positive social change.

Rishabha Malviya • Shivam Rajput

Neurogenetic and Neurodevelopmental Disabilities

Rishabha Malviya
Department of Pharmacy
School of Medical and Allied Sciences
Galgotias University
Greater Noida, Uttar Pradesh, India

Shivam Rajput
Department of Pharmacy
School of Medical and Allied Sciences
Galgotias University
Greater Noida, Uttar Pradesh, India

ISSN 3004-9709 ISSN 3004-9717 (electronic)
SpringerBriefs in Modern Perspectives on Disability Research
ISBN 978-981-95-9222-7 ISBN 978-981-95-9223-4 (eBook)
https://doi.org/10.1007/978-981-95-9223-4

This Springer imprint is published by the registered company Springer Nature Singapore Pte Ltd.
The registered company address is: 152 Beach Road, #21-01/04 Gateway East, Singapore 189721, Singapore

Foreword

In the academic and clinical literature, the book "Neurogenetic and Neurodevelopment disabilities" is an important resource that addresses the urgent need for a comprehensive analysis of the genetic pathways driving neurological growth and disabilities. This book summarises important recent developments in developmental medicine, neurology, and genetics, highlighting differences in genetic and developmental processes that affect behaviour, brain function, and disability.

The book takes a comprehensive, interdisciplinary approach, integrating molecular genetics, nervous system development, and functional outcomes to address the intricacies of neurogenetic and neurodevelopmental impairments, which frequently appear as complex disorders. Its systematic progression from fundamental concepts to specific genetic disorders and related impairments allows for a consistent understanding of the transition from genetic alterations to different clinical symptoms.

The chapters not only explain genetic and neurobiological mechanisms but also emphasise their implications for diagnosis, early intervention, rehabilitation, and long-term care. By addressing genetic and chromosomal syndromes alongside nervous system, motor, muscle, and sensory organ disorders, the book highlights how disruptions at the molecular and developmental levels manifest as functional impairments across the lifespan.

The authors clearly explain basic scientific concepts while highlighting recent advancements in diagnosis and treatment. The consequences of these advancements for early intervention, quality of life, social inclusion, and the need for compassionate, patient-centred treatment are thoroughly explored, reflecting current perspectives in disability studies.

In an era increasingly defined by precision medicine, the book explored the importance of integrating genetic insights into clinical decision-making and rehabilitation planning. It serves as a valuable resource for a broad spectrum of stakeholders, including neurologists, geneticists, paediatricians, rehabilitation professionals, educators, and researchers, and is committed to advancing inclusive care frameworks.

This book, authored by Dr. Rishabha Malviya and his team, provides an in-depth discussion of disabilities through integrated genetic and neurological viewpoints, successfully connecting basic science with real-life applications. The book is positioned to continue being a vital resource for both professionals and up-and-coming researchers in the disciplines of neurogenetics and developmental disorders because of its well-balanced emphasis on fundamental scientific concepts and therapeutic applications. In the end, it is anticipated that this study will promote exhaustive research and interdisciplinary discussion, leading to improved diagnosis, treatment, and assistance for people with neurogenetic and neurodevelopmental disabilities.

Happy reading and best wishes.

Rural Engineering Department
Government of Uttar Pradesh
Vikas Bhawan Ground Floor
Ayodhya, India

Tirankari Mani Tripathi

Preface

The book "Neurogenetic and Neurodevelopmental Disabilities" explores the complex interaction of genetics and multiple scientific disciplines, including neuroscience and developmental biology, in the understanding of disabilities that have a genetic and a neurodevelopmental cause. The book aims to bridge the gap between scientific knowledge and appropriate clinical practices that are evolving and focus on the conversion of findings to effective rehabilitation approaches and models of inclusive care. It provides a comprehensive overview of the biological mechanisms underlying neurogenetic and neurodevelopmental impairments, detailing their clinical manifestations and functional consequences. The first chapter presents basic principles, genetic mechanisms, and classification systems that are necessary in order to understand the progression from molecular changes to functional impairments. Following this chapter, the book goes on to discuss the concept of genetic and chromosomal syndromes, resolving the genotype–phenotype relationships and problems in genetic counselling, which is very important in the care of the patient and family.

The next chapter discusses disorders affecting the nervous system, muscle, and motor pathways, which can lead to diverse disability patterns. They highlight the importance of early detection, clinical evaluation, and therapeutic progress and advances, such as personalised medicine. The final chapter expands its focus to include sensory organ disorders, illustrating the significant effects of visual and auditory impairments on communication, learning, and social engagement, particularly in pivotal developmental periods.

An important objective of this book is to discuss translational approaches linking basic research with the clinical approach to genetic diseases to lead to early diagnosis and multidisciplinary intervention towards an improved quality of life.

Aiming to make the book a resourceful reference and an impetus to further research, interdisciplinary research, and innovation in the field of neurogenetic and neurodevelopmental disabilities, the editors hope that this book will become an extensive resource and contribute to better treatment of affected individuals during their lifetime.

Greater Noida, Uttar Pradesh, India

Rishabha Malviya
Shivam Rajput

Competing Interests The authors have no competing interests to declare that are relevant to the content of this manuscript.

About the Book

The book “Neurogenetic and Neurodevelopmental Disabilities” is a comprehensive academic book that explores the genetic, molecular, neurological, and developmental foundations of disabilities affecting the nervous system, musculature, and sensory organs. With rapid advances in genetics, neurobiology, and clinical diagnostics, doctors, researchers, educators, and clinical experts must understand the genetic and neurodevelopmental causes of disabilities. This multidisciplinary book covers basic science, clinical practice, and rehabilitation.

The book begins by providing a strong conceptual framework in neurogenetic disability, introducing biological, genetic, and developmental pathways that cause functional deficits. It explores the relationship between gene mutations, epigenetic regulation, neurodevelopmental pathways, and environmental interactions and lifespan disability. The chapter emphasised the need and necessity of early diagnosis, genetic counselling, and the role of translational research in improving patient outcomes.

The next chapters explore the genetic and chromosomal syndromes associated with disability, offering in-depth coverage of inherited and sporadic conditions that result in cognitive, behavioural, motor, and sensory impairments. The chapter discusses the clinical aspects, diagnostic criteria, and novel treatments of chromosomal anomalies, single-gene diseases, and complex genetic syndromes.

The book further discusses the nervous system disorders and their contribution to disability, encompassing both central and peripheral nervous system pathologies. They include neurodevelopmental disorders, neurodegenerative diseases, and congenital defects, which inhibit cognition and behaviour. The implications of these diseases on quality of life and functional independence in the long run are pointed out.

A dedicated chapter focuses on muscle and motor system disabilities, exploring neuromuscular disorders, motor neuron diseases, and movement-related impairments. The chapter discusses motor dysfunction's genetic and neurological causes, diagnostic advances, and new rehabilitative and assistive technology to improve mobility and autonomy.

The final chapter addresses sensory organ disorders and associated disabilities, covering visual, auditory, and multisensory impairments, highlighting the highly significant impact of genetics, neurodevelopment, and sensory disabilities on communication, learning, and social integration. New diagnostic tools, intervention strategies, and inclusion strategies are discussed in order to benefit people in different developmental stages.

Overall, the book "Neurogenetic and Neurodevelopmental Disabilities" serves as an excellent resource that integrates genetics, neuroscience, clinical medicine, and rehabilitation science. It is designed to assist students, physicians, geneticists, neuroscientists, special educators, and researchers in becoming familiar with the disability mechanisms, diagnosis, and management. The book contributes to the field of knowledge, collaboration across disciplines, and evidence-informed practice to enhance the outcomes in neurogenetic and neurodevelopmental disabilities significantly.

Contents

Chapter 1
Introduction to Neurogenetic Disability

Abstract Neurological disability, or neuro-disability, a neurological genetic disability, is a condition that involves disorders of the neural system, resulting in functional impairment and any difficulty experienced by individuals in fulfilling the essential and instrumental roles of daily living. This chapter will discuss the various neurogenetic and neurodevelopmental disorders in human beings. It looks at the way genetic and neural developmental disorders affect the brain's ability to work, behaviour, and overall growth. The chapter explains the biological and developmental factors contributing to such disorders and looks at their impact on human abilities. It emphasizes the importance of early diagnosis and effective treatment. The chapter will discuss the fact that neurogenetic disability and families experience enormous social barriers, which consist of a lack of access to inclusive education, healthcare, and employment; social stigma; and emotional distress.

Keywords Neurogenetic · Disorder · Illness · Genetic disability · Accidental disability · Drug abuse

1.1 Introduction

The disability burden related to neurological disorders is high and may have significant implications for patients' autonomy, particularly when they have motor and cognitive impairments. These abnormalities can progressively develop over time after acute disorders such as strokes, severe brain injuries, or spinal cord injuries and many neurodegenerative diseases. The clinical and societal implications of these disorders are high (Baricich et al., 2022). Neurological disability, or neuro-disability, which is defined as impairments of the nervous system resulting in functional limitations, is the impairment of the essential and instrumental activities of daily living of the individuals (Morris et al., 2013; Rana et al., 2025). Neurological disorders may be abruptly displayed, as in the case of spinal cord injuries that occur due to trauma, or they may be progressive, as in the case of multiple sclerosis

R. Malviya, S. Rajput, *Neurogenetic and Neurodevelopmental Disabilities*, SpringerBriefs in Modern Perspectives on Disability Research,
https://doi.org/10.1007/978-981-95-9223-4_1

diagnoses. According to Stancu et al. (Stancu et al., 2025), the major causes of significant disability in the elderly are referred to as neurological diseases, which include those of dementia, stroke, epilepsy, and Parkinson's, which comprise about half of all disabilities among people 65 years of age and above and about 90 per cent of serious dependence in the elderly. Such problems also hamper the social participation of the elderly people, thus affecting their welfare negatively. The elderly who have neurological conditions show very low rates of social participation as compared to other disabilities, including orthopaedic challenges (Pasin & Dogruoz Karatekin, 2024).

Neurological diseases often impose additional impairments on patients, significantly affecting various aspects of their autonomy, including both motor and cognitive abilities. Such anomalies can be progressive and persistent, as in the case of neurodegenerative diseases, or happen as a result of acute events like stroke, traumatic brain injuries, or spinal cord injuries. Neurological issues are also genetic. Hereditary factors can impose numerous limitations on a person. Neurogenetic conditions have been clinically known for over a few decades before the discovery of the first genetic aetiology in 1986, comprising pathogenic dystrophin gene mutations leading to Duchenne muscular dystrophy. Clinical and genetic studies have identified over 1700 monogenic neurological disease causes since that time (Chen et al., 2020). A major concern across all socioeconomic strata is the comprehensive and fair integration of individuals with disabilities. Despite the advances in terms of social awareness related to the significance of inclusion, there are still major barriers that cannot be overcome, as it is believed to influence the quality of life and the overall development of individuals with disabilities (Sandford et al., 2022). Some of the issues affecting this demography are poor access to inclusive education and job discrimination. The absence of clear regulation still alienates the impaired individuals and makes them unable to engage in the society completely (Schalock & Verdugo, n.d.). The disabled face problems in many aspects, mostly the attitudinal, physical environment, transportation, policy, and lack of support by personnel and service providers (Anaby et al., 2013). Disability is not a static and dichotomous process but a dynamic one, which depends on the strengths and weaknesses of an individual or family and the support systems that are available to them (Gómez Sánchez et al., 2021; Thompson et al., 2014). Like everyone else, people with disabilities have the right to live in the community and independently. Hence, interventions, services, and supports should be designed on the basis of teamwork and a thorough understanding of disability that is informed by individual experience and knowledge of experts (Gómez Sánchez et al., 2021; Thompson et al., 2014).

Clinical and social implications of such diseases are enormous. The significant emphasis on immediate treatment has led to insufficient research on the long-term effects of strokes, which significantly contribute to disability. Although stroke is usually viewed as a temporary condition to be treated and disregarded, the majority of stroke survivors continue to experience significant changes in their everyday life, which have a long-term and substantial impact. A half of stroke survivors experience unmet needs, which include incontinence, psychological disorders, mobility problems, pain, and speech problems. However, most of them do not receive rehabilitative follow-up or other forms of therapy (Murray et al., 2003). Recovery is recognized to be a complicated process comprising neurological impairments, probably necessitating a synthesis of spontaneous processes, learning-dependent processes, and adaptive behaviours. Several different mechanisms are suspected to be at play, such as repairing impaired neural tissue functionality (e.g. restitution), rearranging intact neural circuitry (e.g. substitution), improving impaired skill in everyday activity (e.g. compensation), and most importantly, restoring cognitive capabilities (Kwakkel et al., 2004; Elipe-Lorenzo et al., 2025). In this chapter, the numerous neurogenetic and neurodevelopmental illnesses in humans will be addressed. It explores the way genetic and neural developmental abnormalities affect brain functioning, behaviour, and development in general. The chapter has elaborated on neurogenetic and neurodevelopmental disorders' genotypes and phenotypes in a very detailed manner. It looks into the genetic basis of neurodisability, focusing more on diseases caused by chromosomal anomalies and the numerous impairments that they cause. The chapter further discusses neurodisabilities caused by strokes, severe trauma, and other accidental causes that cause impairment of the neurological functions. It further studies the effects of substance and drug abuse on the nervous system with reference to the way such exposures could lead to permanent neurological weaknesses. The chapter evaluates the social, physical, and psychological barriers to the people with neurodisabilities that present an insight into the obstacles people encounter in their daily lives.

1.2 Neurogenetic Disability

Neurodevelopmental disabilities (NDDs) are described as the inability to follow through on cognitive, emotional, and motor developmental levels. The general consensus is that NDDs are linked to the disruption of the intricately coordinated mechanisms that enable brain development. About N3% of children worldwide suffer from neurodevelopmental abnormalities, making them a major public health concern (Gilissen et al., 2014). Their diverse aetiology leads to deficiencies in psychomotor skills, communication, cognition, and adaptive behaviour. Autism spectrum disorder (ASD), epilepsy, attention deficit hyperactivity disorder (ADHD), and intellectual impairment (ID) are all considered NDDs (Tărlungeanu & Novarino, 2018). Numerous investigations have revealed that same genetic pathways could explain the different clinical presentations that characterise NDDs (Hormozdiari et al., 2015). It is therefore common to see two or more of these illnesses co-occurring. For instance, it is common to find individual patients with epilepsy, autistic spectrum condition, and intellectual disability co-occurring (Du et al., 2018). Genetic NDDs have been linked to various forms of mutation, such as chromosome rearrangements, additions and removals of copies, small insertions and deletions (indels), and point mutations. Thus, it is a difficult task to determine a probable underlying mutational occurrence, commonly known as molecular diagnostics, which needs to be mindful of the number of variations that exist in this complex assortment of genomic variations. Despite such challenges, effective genetic counselling and effective patient management require the discovery of genes associated with neurodegenerative disorders. This recognition is an important initial step toward a more profound understanding of the molecular pathways that are influenced by these conditions.

1.2.1 Genotypes and Phenotypes of Neurogenetic Disability

Neurogenetic disabilities exhibit significant variation in their aetiology and clinical features. An example of such a disorder that is classified as a Mendelian monogenic condition is called Rett syndrome and is associated with mutations in the MECP2 gene (Gold et al., 2024). Polygenic diseases, such as Alzheimer's disease, are characterized by the existence of multiple risk variables and lifestyle factors, such as the APOE4 allele (Tripathi et al., n.d.), which play a crucial role in the pathophysiology of the disease (Sun et al., 2024). The effects of neurogenetic diseases may be noticed as early as during the in-utero stage, since they revealed that the foetal activity was altered among patients with type 0 spinal muscular atrophy (Giess et al., 2024). On the other hand, genetic variables have a significant effect on the likelihood of developing dementia and mobility deficits in the adults. Indicatively, in a comparison of cases versus controls, the odds ratio (OR) of glucocerebrosidase (GBA) variation is at 5.4, which subjects one to a higher risk of developing Parkinson disease (Lovras et al., 2025; Dos Santos et al., 2024). Variants of essential genes related to the nervous system may be pathogenic and cause a wide range of phenotypic effects, including intellectual disabilities, autism spectrum disorders, leukodystrophies, movement disorders, psychiatric disorders, and dementias. These abnormalities are the result of a diverse array of genetic variants, including minor insertion or deletion alleles, larger copy number variants (CNVs), single-nucleotide variants (SNVs), and whole-chromosome aneuploidy, as occurs in Down syndrome (Fu et al., 2022). While many mutations result in impairment or functional loss of an encoded protein, other mutations are harmful because they induce a gain of function, such as an ion channel remaining open for a long time. Additional pathogenetic pathways include epigenetic changes (Reichard & Zimmer-Bensch, 2021) and expansions of toxic repeats (Zhou et al., 2022; Devinsky et al., 2025).

1.2.2 Genetics of Neurogenetic Disabilities

The identification of probable hereditary origins is necessary to comprehend the molecular mechanisms that underlie the onset of NDDs. Additionally, it makes it easier to identify genotype-phenotype associations, which could help track the disorder's development and predict potential problems. Even though many genes linked to NDDs have been identified, many people with these problems still do not have a molecular diagnosis. Additionally, research on the relationship between genotype and phenotype has shown that people with overlapping genetic aetiology can vary greatly in the number and severity of clinical manifestations (Casanova et al., 2018). As a result, the lack of heredity and the variation in phenotypic expression suggest that NDDs are multifactorial and/or polygenic. For the analysis of the roles of both genetic and nongenetic variables in the aetiology of NDDs, familial disorders provide a useful framework, especially when there is a shared genetic background. Consequently, a plethora of research has been conducted on either monozygotic twin with discordant phenotypes (Huang et al., 2019; Radley et al., 2019) or on pedigrees with partial penetrance and phenotypic diversity across multiple affected progenies (Karaca et al., 2018). Not only does this area of research hold great promise for determining the disease's molecular cause, but it also holds promise for unravelling risk and protective variables. Additionally, it possesses the capability to help generate more precise genotype-phenotype associations. Genetic susceptibility and mutational burden are the two main principles that have been found to be fundamentally central to the phenotypic result through the analysis of inherited NDDs.

Gene vulnerability is the capacity of a gene to withstand disruptive variations; a higher level of vulnerability is associated with a lesser tolerance to mutations. A significant sensitivity to gene dosage is a characteristic of haploinsufficiency, which is present in some genes associated with neurodevelopmental disorders. These specific genes are classed as being extremely susceptible and there is a considerable risk of disease associated with mutations that impact these genes. Notable instances of genes with elevated susceptibility include DEPDC5, CACNA1A, and SCN8A. Even in the absence of other contributing variables, the disruption of any one of these genes is likely to trigger the manifestation of a disease symptom, resulting in monogenic types of neurodegenerative illnesses. As a result, mutations affecting these genes usually face strong negative selection pressure. Therefore, population studies have reported a lower occurrence of disruptive variants in sensitive genes than at other genomic locations (Iossifov et al., 2015). Mutations that exhibit a high degree of susceptibility are considered rare alterations that have a high penetrance and are linked to a large risk of sickness.

The genes that show a decreased sensitivity to disruptive mutations are located at the opposite end of the vulnerability spectrum. Variants in these genes are frequently transmitted down through families over several generations and are not subject to negative selection pressure (Niemi et al., 2018). These variations are classified as frequent variations that are associated with a low risk of disease, as individual disruptive events that affect nonvulnerable genes do not necessarily cause disease. Recent research indicates that common genetic variants account for a considerable fraction of polygenic NDDs (Kurki et al., 2019). A disease phenotype may appear as a result of the cumulative consequences of various mutational events (Pizzo et al., 2019). The phenotypic outcome is frequently influenced not only by the cumulative effect of individual mutations but also by the physical and/or functional interactions between the implicated genes (i.e. epistasis) (Iyer et al., 2018). The idea of mutational burden, which holds that the number of disruptive events influences the penetrance and complexity of a disease manifestation, is strongly correlated with epistatic interactions and dose sensitivity. For example, a variety of clinical symptoms, such as mobility problems, intellectual disability, autism spectrum disorder, and benign familial infantile seizures, are often associated with loss-of-function monoallelic mutations in the sodium channels CACNA1A and SCN8A (Wengert et al., 2019). According to the previously mentioned criteria for mutational load and dosage sensitivity, inherited germline biallelic mutations of SCN8A and CACNA1A are associated with more severe symptoms than monoallelic mutations. While the heterozygous parents and siblings only exhibit modest cognitive impairment in the absence of seizures, the recently reported compound heterozygous pro bands of CACNA1A and SCN8A are differentiated by the onset of epileptic encephalopathy (Reinson et al., 2016).

In certain instances, a two-hit model a combination of germline and somatic events—can lead to a higher mutational impact. The standard two-hit hypothesis states that a predisposed genetic background is established by a constitutive hereditary mutation. The establishment of a disease phenotype or the aggravation of pre-existing clinical characteristics is then linked to a somatic mutation that happens later in development. Mutations affecting DEPDC5 provide a relevant example of the two-hit model. DEPDC5 germline heterozygous loss-of-function mutations also play a major role in causing familial refractory focal epilepsies (Ribierre et al., 2018). Furthermore, it has been discovered that a subsequent somatic mutation that causes biallelic inactivation of DEPDC5 contributes to the additional emergence of localised cortical dysplasia in patients with a severe phenotype (Lee et al., 2019). The two-hit model could be extended by the fact that there might be primary and secondary variations that occur at other genomic locations. The relevance of different molecular diagnostics with regard to NDDs has been demonstrated by the numerous studies (Liu et al., 2019). The theory of mutational burden is reinforced by the study of genotype-phenotype interactions that has shown that those with mutations in many genes are more vulnerable to being impacted (Guo et al., 2018). Moreover, the occurrence and the severity of the presented clinical manifestations are positively linked with the count of disruptive episodes (Posey et al., 2017). For example, one recent investigation looked at the role of various mutational events in two families with varied intra-family clinical heterogeneity. Additional clinical features in the probands were explained by mutations at other loci that were absent in the less affected siblings in both families. The frequency of common genetic variations may be the initial factor that increases the susceptibility of the genetic background to subsequent disease events. Genetic susceptibility to lower educational level and intellectual disability is strongly connected with the burden of common variation accumulation in families with a history of neurodevelopmental issues, according to recent studies (Rana et al., 2025).

1.3 Disability Due to Chromosome Disorders

According to estimates, 3% of babies are born with a serious structural defect that could lower their quality of life. It is interesting that 0.7% of neonates have numerous substantial anatomic defects, however most of these individuals appear with a single deformity. By the age of five, a major anomaly is detected at an additional rate of 2%. The timely diagnosis of a condition's genetic foundation can aid in therapy and the identification of resources required to provide impacted persons with better health care. The following section will look at genetic illnesses and syndromes marked by underlying chromosomal abnormalities. The application of next-generation sequencing and microarray-based diagnostic methods in genetic evaluation and diagnosis will also be covered. The tightly packed DNA that makes up chromosomes is held attached by proteins and histones. Under a light microscope, they appear as two arms connected by a centromere, each of which has a telomere at the end. First seen in 1956, human chromosomes exhibit unique size, centromeric location, and staining patterns following dye treatment. There are usually 46 chromosomes, 22 autosome pairs, and one set of sex chromosomes (Haldeman-Englert et al., 2018), and each chromosome is recognised by a number. During mitosis, when chromosomes condense and become visible, cells can be examined using a technique called karyotype analysis. Amniocytes, skin fibroblasts, and peripheral blood lymphocytes cells are often utilized for the examination. Additionally, chorionic villi and bone marrow cells can be successfully karyotyped. High-resolution analysis of up to 800 bands for structural rearrangements and resolution of at least 400 bands across all chromosomes are made possible by the widely used cytogenetic technique known as G-banding.

In order to generate gametes, homologous chromosomes must align and cross over during meiosis, which exchanges genetic material. After recombination and reduction division in meiosis I, the cell's diploid composition is reduced to 23 chromosomes as the recombined pairs split apart. Chromosomes split during meiosis II, and fertilization restores the complete diploid state (Nussbaum et al., 2007). Aneuploidy is the term used to describe an imbalance of genetic material caused by a net gain or loss of genetic material during embryonic divisions or sperm or egg development. Trisomy (three copies of a whole chromosome) and monosomy (one copy of a complete chromosome) are examples of classic aneuploidy syndromes. Trisomies can arise from nondisjunction, a failure of normal chromosome separation that occurs during egg development, particularly in maternal meiosis I. This nondisjunction is particularly noticeable in XXX trisomy (MacDonald et al., 1994) and acrocentric chromosomal trisomies. Because meiotic nondisjunction is more common in older mothers, women 35 years of age and older may have chorionic villus sampling (CVS) or amniocentesis for prenatal karyotyping (Hook & Cross, 1982). A new technique called noninvasive prenatal screening determines the copy number of particular chromosomes by analysing the cell-free foetal DNA fraction that circulates in the mother's peripheral blood (Hardisty & Vora, 2014). Nondisjunction, which occurs during mitosis and results in an uneven division of genetic material during early embryonic cell division, can result in the production of two viable cell lines. This can be caused by sex chromosomal aneuploidies, including mosaic Turner syndrome. When nonhomologous chromosomes rearrange in balanced translocation carriers, partial aneuploidy may occur, increasing the likelihood of unbalanced rearrangements in progeny. Deletion syndromes are caused by the loss of genetic material from several chromosomes and display a unique phenotype (Emanuel & Shaikh, 2001). Large DNA blocks called segmental duplications, which include repeating sequences unique to a chromosome, are linked to abnormal recombination and other chromosome microdeletions. Segmental duplications in regions prone to rearrangements result in phenotypes such as Williams-Beuren syndrome, Prader-Willi or Angelman syndrome, Charcot-Marie-Tooth conditions, hereditary neuropathy with a risk of pressure palsies, and recurrent 22q11.2 deletion syndromes (Emanuel & Saitta, 2007). Smaller chromosomal rearrangements that are frequently missed by conventional karyotyping methods can now be identified due to the use of chromosomal microarrays. Clinical studies have further defined the related phenotype, and high-resolution microarrays in newborns with various congenital abnormalities have frequently resulted in the identification of a particular genotype (Bejjani & Shaffer, 2008). Figure 1.1 and Table 1.1 illustrates various major neurogenetic disabilities due to chromosome disorders.

1.3.1 Down Syndrome (Trisomy 21)

According to 1960 research by Lejeune and Turpin, Down syndrome affects roughly 1 in 700–800 live births and is a prevalent autosomal trisomy (Hook, 1992). Meiotic nondisjunction accounts for more than 90% of instances, with a notable maternal age effect. Three to five per cent of cases are caused by a translocation, which can be de novo or inherited from a balanced parent. When a translocated chromosome rearranges with another acrocentric chromosome, this is known as a Robertsonian

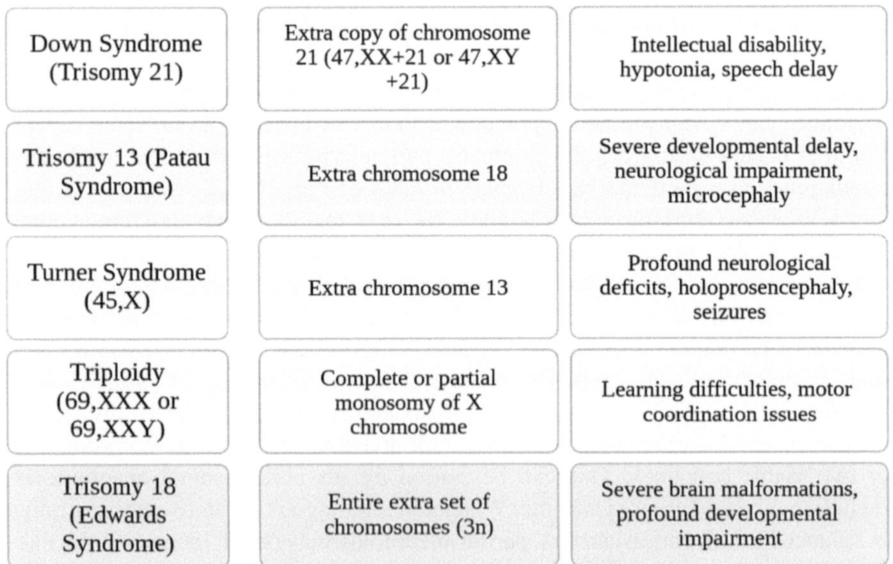

Fig. 1.1 Major neurogenetic disabilities due to chromosome disorders

Table 1.1 Major neurogenetic disabilities due to chromosome disorders

Condition	Genetic basis	Key neurodevelopmental features	Severity
Down syndrome (trisomy 21)	Extra copy of chromosome 21 (47, XX + 21 or 47, XY + 21)	Intellectual disability (mild–moderate), delayed milestones, hypotonia, speech delay, increased risk of Alzheimer-like changes	Moderate
Trisomy 18 (Edwards syndrome)	Extra chromosome 18	Severe developmental delay, neurological impairment, microcephaly, poor muscle tone	Severe, often life-limiting
Trisomy 13 (Patau syndrome)	Extra chromosome 13	Profound neurological deficits, holoprosencephaly, seizures, severe cognitive impairment	Very severe, high neonatal mortality
Turner syndrome (45, X)	Complete or partial monosomy of X chromosome	Learning difficulties (esp. spatial), motor coordination issues, normal intelligence usually preserved	Mild
Triploidy (69, XXX or 69, XXY)	Entire extra set of chromosomes (3n)	Severe brain malformations, profound developmental impairment, incompatible with long-term survival	Very severe, typically lethal

translocation. In 3% of cases, there is also mosaic Down syndrome, commonly known as mitotic nondisjunction. The phenotypic characteristics of Down syndrome, which include brachycephaly, a third or confluent fontanelle, and other anomalies, usually make the disorder identifiable at birth. However, the diagnosis could be more challenging due to ethnic variations or prematurity. The diagnosis and its cause are confirmed by an instantaneous karyotype, which is also utilized for recurrence risk counselling. Down syndrome can cause problems in several organ systems, including tetralogy of Fallot, patent ductus arteriosus, ventricular septal defects, atrial septal defects, and atrioventricular tricular canal defects. Medical and surgical procedures are common, and an echocardiography is necessary in every situation. Malformations of the digestive system have also been reported, including duodenal atresia, Hirschsprung disease, and other less common diseases. Monitor the baby's meals and bowel motions before releasing them from the nursery (Bull, 2011). Patients with Down syndrome acquire distinct growth curves as a result of their markedly reduced postnatal growth velocity. According to health supervision standards, growth curves from the Centres for Disease Control and Prevention or the World Health Organization should be used to evaluate patients. Because of prevalent disorders such glaucoma, cataracts, myopia, and strabismus, initial ophthalmologic assessments are advised in the first few months of birth and once a year. Middle ear disease contributes to heterogeneous hearing loss, which affects approximately half of individuals (Fernandes et al., 2001).

Neurologic issues might result from spinal cord compression, so doctors should be careful while assessing the cervical spine, particularly prior to anaesthesia. Five per cent of individuals have hypothyroidism, which is frequently accompanied by thyroid autoantibodies. Newborn screening programs are part of the first assessment, and then there are follow-up measurements at six, twelve, and yearly intervals. Since bone marrow dyscrasias and transient self-resolving myeloproliferative disorders are common in the first year of life, a complete blood count with differential should be performed at birth. Until the age of sixteen, leukaemia is more common, and congenital or neonatal cases of acute nonlymphoblastic leukaemia are more common. Additionally, there is a higher risk of iron-deficiency anaemia, and it is advised to evaluate haemoglobin levels annually (Shumate et al., 2025). Individuals with Down syndrome have a range of behavioural traits, personalities, and developmental capacities. Early intervention and therapy are necessary since central hypotonia and motor delay are most noticeable during the first 3 years of life. There are contradictory findings regarding hereditary and environmental factors that influence intelligence. Five to ten per cent of patients have seizure disorders, frequently during infancy (Pueschel et al., 1991). Patients with Down syndrome frequently pass away from cancer, infections, and congenital heart problems. The long-term survival rate is good after congenital abnormalities are corrected, but less than half live to 60 years and less than 15% live over 68 years (Fong & Brodeur, 1987). Overexpression of the amyloid precursor protein gene causes early-onset beta-amyloid plaques in the brain, which are common in patients over 40 with neurodegenerative illnesses that resemble Alzheimer's. Other symptoms include neurone loss, oxidative damage, neuroinflammation, neurofibrillary tangles, white matter disease, and cerebrovascular pathology (Head et al., 2016). While a tiny percentage of women with Down syndrome have had children, men with the condition are nearly always infertile.

1.3.2 Trisomy 18 (Edwards Syndrome)

One out of every 6000 live babies have trisomy 18, a rare condition that has a significant intrauterine death rate. Only five per cent of conceptuses survive to birth, and thirty per cent of foetuses detected by second-trimester amniocentesis die before the end of pregnancy (Tosun et al., 2025). Prenatal ultrasound may reveal the condition, and when results are suggestive, definitive testing is advised. This disorder is characterized by phenotypic features such intrauterine growth restriction, a short, narrow cranium, open metopic suture, low-set ears, micrognathia, clenched hands, hypoplastic nails, and "rocker-bottom" feet. Other deformities include congenital heart disease, clubfoot deformity, cleft palate, neural abnormalities, choanal atresia, ocular abnormalities, vertebral abnormalities, hypospadias, cryptorchidism, and limb issues. Only 5% of infants with this condition are still alive at 1 year of age, and almost 90% of them pass away during the first 6 months (Armour et al., 2025). Congestive heart failure, infection, and central apnoea are among the potential causes. Poor growth and nutrition are common throughout the neonatal era, and some survivors develop malignant tumours. During prenatal or postnatal visits, parents should be informed about interventions that can enhance survival and quality of life (Srinivasan et al., 2025; Kosho & Carey, 2016).

1.3.3 Trisomy 13 (Patau Syndrome)

A rare genetic disorder, trisomy 13 affects 2% to 3% of foetuses and occurs in 1 in 12,500 to 1 in 21,000 live births. Prenatal diagnosis by CVS or amniocentesis may be prompted by abnormal prenatal screening or foetal ultrasonographic results. Midline malformations associated with trisomy 13 include kidney abnormalities, holoprosencephaly, congenital heart disease, cleft palate, and postaxial polydactyly. The diagnosis consists of scalp malformations, ocular abnormalities, and microcephaly. More than half of people with seizure disorders have brain abnormalities such as holoprosencephaly. In half of instances, the disease is associated with a scalp deformity called cutis aplasia. Eighty per cent of patients have congenital cardiac disease, which frequently leads to dextrocardia, VSD, ASD, or PDA. There are also limb abnormalities such as hyperconvex narrow fingernails, single palmar creases, and postaxial polydactyly. Similar to Barr bodies, neutrophils exhibit a higher frequency of nuclear projections, especially in men. A foetus with trisomy 13 has a terrible prognosis; less than 5% survive to 6 months of age, and neonatal mortality is 80%. Deafness, blindness, and mental retardation are prevalent. Feeding issues are frequent. Interventions can improve quality of life and increase survival, but parents should talk about these options (Tran et al., 2023; Anacleto et al., 2024).

1.3.4 45, X (Turner Syndrome)

One in every 2500 infants has Turner syndrome, a condition in which a female conceptus loses one X chromosome. Half of cases are caused by the 45 X karyotype, with the remaining cases being caused by other X chromosomal abnormalities. Over 99% of foetuses with a 45 X complement spontaneously terminate, whereas only 0.1% make it to term. Studies show that in 80% of cases, the paternally derived X chromosome is removed (Stubbins et al., 2021). The phenotypic changes that characterize Turner syndrome include short height, webbed neck, craniofacial deviations, hearing loss, shield chest, renal anomalies, lymphoedema, and congenital heart disease. Coarctation, mitral valve prolapses, bicuspid aortic valve, and valvular aortic stenosis are common abnormalities. Patients with Turner syndrome frequently experience growth problems, especially short height. With an initiation age as early as 9 months, growth hormone therapy can improve adult height (Gravholt et al., 2023). Gonadal dysplasia-induced primary ovarian failure can postpone primary amenorrhoea and secondary sexual traits. Menstruation, bone mass, and secondary sex traits are all aided by cyclic hormonal therapy (Steiner & Saenger, 2022). Donor oocytes and assisted reproduction methods can be used to treat infertility, which is a prevalent condition. Prior to pregnancy, it is essential to assess individuals for structural cardiovascular abnormalities. The syndrome, which is distinguished by challenges in spatial and perceptual cognition, results in a reduced performance intelligence quotient.

1.3.5 Triploidy (69, XXX or 69, XXY)

The term "triploidy" refers to a karyotype where each chromosome is found in three copies. This condition is caused by the fertilization of the ovum by two distinct sperm (dispermy) and the total lack of normal chromosomal segregation during maternal meiosis. Up to 15% of pregnancy losses with chromosome defects are spontaneous terminations, which account for the majority of triploid foetuses. Affected foetuses rarely survive past infancy, and anecdotal evidence of survival is available. There have been reports of mosaicism involving diploid and triploid cell combinations (mixoploid). Associated findings include anomalies such as hydrocephalus, neural tube malformations, ophthalmic and auricular abnormalities, cardiac problems, and three to four finger syndactyly. Furthermore, the placenta is frequently bigger, irregular, and cystic (Rogol, 2023; Haldeman-Englert et al., 2018).

1.4 Neurogenetic Disability from Accidents

1.4.1 Stroke

Stroke, the second most prevalent cause of death globally, underscores the necessity of enhancing the rapid treatment of the condition in order to substantially reduce mortality rates (Powers et al., 2018). Although strokes are a leading cause of neurodisability, the long-term effects have not been thoroughly studied due to this primary focus. Even though strokes are usually thought of and treated as a temporary illness, the majority of stroke survivors suffer from ongoing, significant deficits in their capacity to perform daily chores. According to estimates, 50% of stroke survivors have missed opportunities related to speech, mobility, pain, mental problems, and incontinence. However, most people do not obtain a rehabilitative follow-up or other treatment techniques (Murray et al., 2003). The process of recovery is recognized as a complex phenomenon that most likely combines learning-based mechanisms, adaptive behaviours, and spontaneous healing. According to recent research, several factors contribute to this process, including the repair of damaged brain tissue (restitution), the realignment of preserved brain connections (substitution), the enhancement of functional abilities through alternative means (compensation), and most significantly, the recovery of cognitive abilities (Kwakkel et al., 2004). An increasing body of research suggests that interdisciplinary rehabilitation is advantageous for stroke survivors, particularly when implemented during the acute and subacute phases of the event (Dobkin, 2005; Langhorne et al., 2011). After a stroke, systematic therapy for motor function is often considered to be completed in three to 4 months. The basic concept of this assessment is that motor and functional recovery usually takes 3 to 6 months after the stroke occurrence (Baricich & Carda, 2018). Recent studies, however, suggest that motor and cognitive abilities may continue to improve for a long time (Rohrbach et al.; Wang et al.). This potential for ongoing recovery has also been associated with a variety of neurological disorders that affect the central nervous system and hereditary muscular diseases (Baricich et al., 2022).

1.4.2 *Traumatic Brain Injury (TBI)*

Traumatic brain injury (TBI) can range in severity from mild consciousness loss to death and extreme coma. Treatment options vary from cognitive therapy to major surgery, depending on the severity of the injury. The best management practices are situation-specific and cannot be applied to every person. Millions of people are impacted by TBI each year, and between 2001 and 2010, the number of ED visits, hospital stays, and fatalities increased (Dong et al., 2025). TBI-related mortality has, however, declined as a result of improved awareness, advancement in management, and innovations in technology. Overall, TBI rates are underreported since a portion of cases never receive medical attention (Masel et al., 2014). The highest rates are found in young people (0–4 years old), adolescents and young adults (15–24 years old), and the elderly (>65 years old). TBI is most commonly caused by motor vehicle accidents and falls (Rutland-Brown et al., 2006). As a result, the number of people with severe disabilities that are directly linked to the TBI is increasing. TBI can cause either short-term or long-term neurological deficits, and its complex aetiology includes both primary and secondary damage. While secondary damage happens from minutes to days and includes a chain reaction of molecules, chemicals, and inflammation, primary deficiencies are directly linked to the external impact on the brain. Apoptosis, elevated intracellular calcium, and neuron depolarization all contribute to this process, which breaks down cells and damages the blood-brain barrier. The mediators upregulate and downregulate the process. Anatomical, molecular, and functional reorganization occur during the recovery phase following the second damage. Brain parenchyma makes up 83% of the intracranial compartment, followed by cerebrospinal fluid (11%) and blood (6%). When volume is more than usual, compensatory mechanisms such as venous congestion, cytotoxic and vasogenic oedema, and mass effects from blood take place. The incompressibility of brain tissue causes blood to extrude from the brain and CSF to extrude to the spinal compartment. These compensatory systems malfunction in the absence of appropriate intervention, leading to severe brain compression and death (Greenberg, 2006). Concussions are frequently regarded as mild traumatic brain injuries (TBIs) that do not cause substantial structural damage (Piedade et al., 2021). They usually appear as a consequence of acceleration/deceleration forces and head trauma. A temporary change in mental status, from bewilderment to unconsciousness, may result from the damage. Routine neuroradiographic imaging, such as CT scans and MRIs, does not reveal any evident abnormalities. Newer methods, nevertheless, might result in an earlier diagnosis. It has been proposed that even minor TBIs may result in mild axonal damage (Akira et al., 2022). A rare disease known as “second impact syndrome” occurs when athletes have a second concussion while recuperating from their first injury. After a brief period of time on the pitch, this condition frequently leads to malignant cerebral oedema, which has a 50% to 100% death rate.

1.4.3 Chronic Traumatic Encephalopathy (CTE)

CTE is a delayed appearance that can arise from repeated mild traumatic brain injury. This entity has received a lot of attention because one of the challenging effects of CTE is the development of mental health issues, which can lead to suicidal behaviour in individuals. Other CTE symptoms include dysarthric speech, tremors, concentration problems, memory and executive function impairments, incoordination, and pyramidal signs. CTE is most likely a result of the progressive degeneration of neuronal structures (Abad et al., 2022; Papini et al., 2025).

1.4.4 Extra-Axial Hematomas

Subdural haematomas (SDH) and epidural haematomas (EDH) are both considered extra-axial haematomas. The most common cause of extradural haemorrhages is a direct blow to the temporal region, which can sometimes result in a fractured skull and the destruction of the middle meningeal artery. However, venous injuries, such as displacement of the transverse sinus, have also been associated with a higher frequency of posteriorly directed epidural haemorrhages. A rapidly EDH may result in a person with largely normal cognitive function, which may then deteriorate as herniation syndromes develop once the patient's intracranial pressure (ICP) reaches a crucial threshold (Gok et al., 2023).

Epidural haemorrhages are typically observed in acute medical conditions, while subdural haemorrhages can manifest in a variety of ways, depending on the patient's age and the duration of blood component retention. One typical source of SDH in trauma conditions is the acceleration and deceleration of the brain's surface against the inner surface of the skull, which causes shearing damage to the bridging veins (Madhok et al., 2022). Compared to epidural haematomas, acute subdural haematomas usually include a significantly larger degree of underlying brain injury, making them potentially dangerous for patients in traumatic situations. If untreated, underlying cerebral oedema frequently plays a major role in the development of herniation symptoms and the subsequent midline displacement of structures. The patient may continue to be unaware of some social determinants of health for a long time. Acute and chronic subdural haemorrhages can be observed as the acute blood in the subdural space gradually liquefies. When compared to the acute phase, the clinical picture of subacute and chronic subdural haemorrhages is marked by a noticeably slower and less progressive course. Elderly people are more likely to have chronic-looking subdural haemorrhages, which are frequently seen in combination with antiplatelet or anticoagulant medication. The most common symptoms of subacute and chronic subdural haemorrhages include headaches, hemiparesis, difficulty speaking, disorientation, and changes in mental status. These haemorrhages typically present with a more subtle beginning. Untreated unilateral or bilateral chronic subdural haemorrhage can occasionally cause a patient to show significant neurological deterioration.

1.4.5 Contusions and Traumatic Subarachnoid Haemorrhage

The forces of coup and contrecoup usually result in contusions. Coup injuries occur at the site of impact, but contrecoup injuries usually occur on the other side of the hit and frequently result in damage to the frontal and anterior temporal lobes. Trauma is the main cause of subarachnoid haemorrhage, which happens when tiny capillaries burst, and blood temporarily leaks into the subarachnoid area. It is commonly believed that spontaneous aneurysmal subarachnoid haemorrhage (Yue et al., 2024), which happens when blood is released into the subarachnoid space under arterial pressure, is more severe than traumatic subarachnoid haemorrhage.

1.4.6 Diffuse Axonal Injury (DAI)

Diffuse axonal injury (DAI) is the most severe type of axonal shearing injury. Significant rotational acceleration and deceleration forces are typically responsible for the development of such an injury (Mesfin et al., 2025). The brainstem, thalamus, internal capsule, corpus callosum, and corona radiata are among the regions where it is frequently observed radiographically as faint haemorrhagic foci on T2 and gradient echo sequences (Mosleh et al., 2024). Depending on the precise location of axonal shearing, patients may present with a wide range of clinical symptoms. Some people with diffuse axonal damage may exhibit altered consciousness for a few days, while others may present with hemiparesis as a result of internal capsule involvement. Axonal integrity issues in particular areas of the reticular activating system cause some people to lose consciousness (Rajput et al., 2025). Figure 1.2 and Table 1.2 describe various acquired neurodisabilities due to cerebrovascular and traumatic brain conditions.

1.5 Neurodisability Due to Substance Abuse

Drug addictions are long-term brain illnesses that impact motivation, reward, memory, and judgement. Under adverse circumstances, these modifications result in challenging behaviours, egocentric activities, and enduring behaviours. In a wide range of drug addicts, repeated exposure to various substances causes cognitive abnormalities which

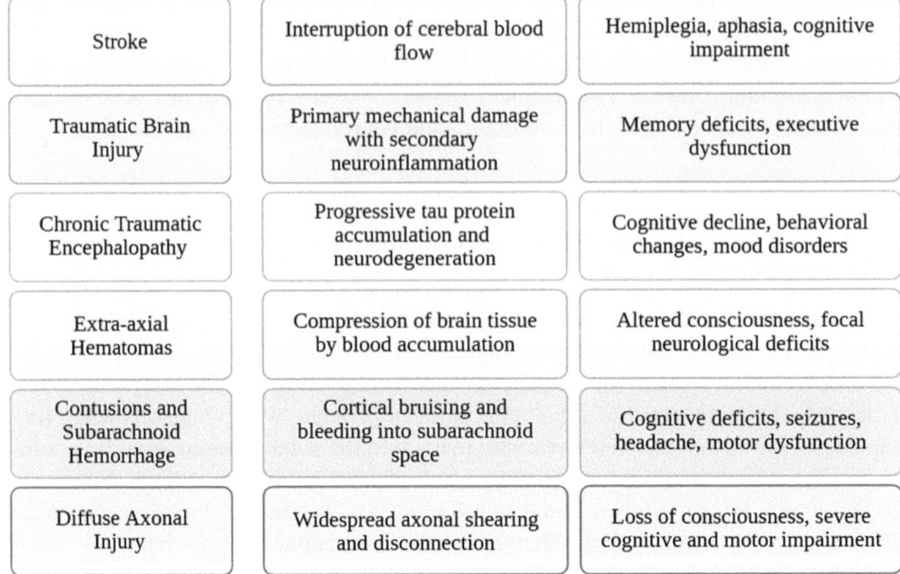

Fig. 1.2 Acquired neuro-disabilities due to cerebrovascular and traumatic brain conditions

Table 1.2 Acquired neuro-disabilities due to cerebrovascular and traumatic brain conditions

Condition	Primary cause	Pathophysiology	Key neurological disabilities
Stroke	Ischemic or haemorrhagic cerebrovascular event	Interruption of cerebral blood flow causing neuronal death	Hemiplegia, aphasia, cognitive impairment, sensory loss
Traumatic brain injury (TBI)	External mechanical force (falls, accidents, assaults)	Primary mechanical damage with secondary neuroinflammation	Memory deficits, executive dysfunction, motor impairment
Chronic traumatic encephalopathy (CTE)	Repetitive head trauma	Progressive tau protein accumulation and neurodegeneration	Cognitive decline, behavioural changes, and mood disorders
Extra-axial hematomas	Head trauma causing epidural or subdural bleeding	Compression of brain tissue by blood accumulation	Altered consciousness, focal neurological deficits
Contusions and traumatic subarachnoid haemorrhage	Blunt head injury	Cortical bruising and bleeding into the subarachnoid space	Cognitive deficits, seizures, headache, and motor dysfunction
Diffuse axonal injury (DAI)	High-velocity acceleration-deceleration injury	Widespread axonal shearing and disconnection	Loss of consciousness, severe cognitive and motor impairment

can be diagnosed by MRI as evidence of brain dysfunctions. These actions have the potential to be fatal and are linked to particular cognitive disorders. It is challenging to assess the incidence of addictions, which are brain illnesses that impact reward, motivation, and memory, because of illicit substances including cocaine, opiates, amphetamines, and cannabis. The pharmacological agent used such as cocaine, marijuana, and methamphetamine determine the type of drug addiction. It may result in euphoria, psychological dependence, and an insatiable desire to have the same sensation again. Physical dependency is another side effect of frequent opiate usage. Addiction to these drugs has negative social, psychological, and bodily effects. Structural changes in learning and reward system-related brain regions affect the long-term clinical signs of drug addiction. Mental health issues may result from the biological impacts of addictions on the central nervous system, which alter the composition and functionality of the brain (Volkow et al., 2012). Drug-induced CNS functions have been demonstrated by clinical trials and cognitive testing, and researchers have seen significant changes in addicts' brains (Daglish & Nutt, 2003). Figure 1.3 and Table 1.3 illustrates various neuro-disability associated with substance abuse.

1.5.1 Marijuana (Cannabis)

Cannabis has surpassed amphetamines, opioids, and cocaine as the most commonly produced and used illegal substance in the world. The plant known as Cannabis sativa, which is indigenous to both temperate and tropical climates, has 489 distinct

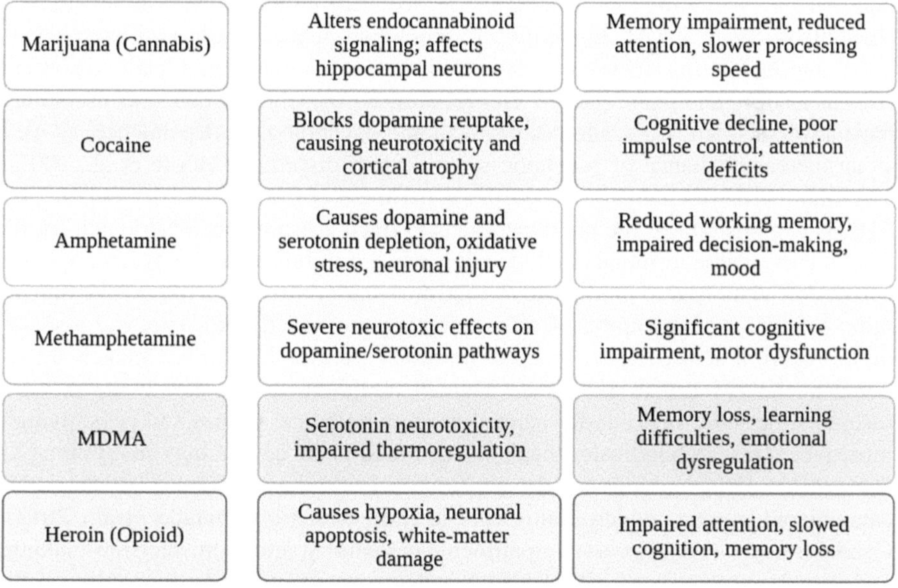

Fig. 1.3 Neuro-disability associated with substance abuse

Table 1.3 Neuro-disability associated with substance abuse

Substance/disorder	Substance type	Neurobiological impact	Key neuro-disabilities
Marijuana (cannabis) use disorder	Cannabinoid	Alters endocannabinoid signalling; affects hippocampal neurons and synaptic plasticity	Memory impairment, reduced attention, slower processing speed, impaired executive function
Cocaine use disorder	Stimulant	Blocks dopamine reuptake, causing neurotoxicity and cortical atrophy	Cognitive decline, poor impulse control, attention deficits, increased stroke risk
Amphetamine analogue use disorders	Stimulant	Causes dopamine and serotonin depletion, oxidative stress, and neuronal injury	Reduced working memory, impaired decision-making, mood and behavioural dysregulation
Methamphetamine use disorder	Strong stimulant	Severe neurotoxic effects on dopamine/serotonin pathways, structural brain damage	Significant cognitive impairment, motor dysfunction, emotional instability, risk of neurodegeneration
MDMA (ecstasy) use disorder	Empathogen–stimulant	Serotonin neurotoxicity, impaired thermoregulation, hippocampal damage	Memory loss, learning difficulties, emotional dysregulation, and sleep disturbance
Heroin (opioid) use disorder	Opioid	Causes hypoxia, neuronal apoptosis, and white-matter damage	Impaired attention, slowed cognition, memory loss, increased risk of hypoxic brain injury

compounds, including 70 cannabinoids like Δ9-tetrahydrocannabinol (THC) (ElSohly & Slade, 2005). In addition to producing pleasure, self-assurance, relaxation, and well-being, its use can result in behavioural and mental health disorders such as anxiety and panic (Hall & Degenhardt, 2009). Chronic cannabis users may have adverse side effects, such as physical and psychological dependence, as well as an increased chance of psychotic symptoms or disorders (Moore et al., 2007). The stimulation of cannabinoid receptors, particularly CB1 and CB2 receptors, by THC is responsible for the pharmacological effects of cannabis (Maldonado et al., 2011). Presynaptic terminals include CB1, which inhibits adenylyl cyclase activity and cAMP synthesis to modify neurotransmission (Glass et al., 1997). The human brain contains a large number of CB1 receptors, particularly in frontal and associational function areas. Higher receptor densities are found in limbic areas and thalamic nuclei (Munro et al., 1993). Human peripheral tissue also contains CB1 receptors, but CB2 is primarily located in immunological tissues and cells, though more recent research indicates that it is also found in the central nervous system (Xi et al., 2011). Drug cravings and drug-seeking behaviour are influenced by the endocannabinoid system, which contributes to drug addiction (Batalla et al., 2013). Cannabis users have cognitive impairments in memory, attention, decision-making, and psychomotor speed. These impairments may influence risk-taking and decision-making, which could have detrimental effects on one's health (Schuster et al., 2012).

Researchers have discovered that marijuana can alter the amount, shape, and makeup of brain tissue, particularly the cerebellum and hippocampal regions (Yücel et al., 2008). These structural alterations might be connected to the reductions in neurones and synapses that THC causes in animal models (Scopetti et al., 2023). Additionally, there is decreased grey matter density in the right parahippocampal gyrus, increased density bilaterally near the precentral gyrus and right thalamus, and decreased grey matter volume in the right anterior hippocampus in marijuana users (Brassard et al., 2025). Furthermore, the prefrontal cortex's volume varies according to gender in heavy cannabis users (Allick et al., 2021). The discovery of cannabinoid receptors in oligodendroglial cells prompted a recent investigation of the white matter state of marijuana users (Fleisher-Berkovich et al., 2023). It has been demonstrated that cannabinoid receptors are present in the white matter areas of the rat foetus and postnatal brain, including the corpus callosum, anterior commissure, fornix, stria terminalis, and stria medullaris (Evanski et al., 2024). As people age, fewer cannabinoid receptors are found in these structures. This suggests that during a developmental period, white matter structures may be susceptible to the effects of exogenous cannabis exposure. The age at which regular cannabis use begins may influence the extent of any white matter alteration in marijuana users (Marinho, n.d.). The axonal connection was disrupted in the right fimbria of the hippocampus, the splenium of the corpus callosum, and the commissural fibres that extend to the precuneus (Gunatilake et al., 2025). According to neuroimaging research, marijuana abuse disorder is associated with structural and functional alterations in several parts of the brain.

1.5.2 Cocaine

According to the UNODC, cocaine, a highly addictive narcotic with a short half-life and strong reinforcing characteristics, is used by 14.2–200.5 million individuals yearly, or 0.3–0.5% of the world's population aged 15–64. Due to its suppression of dopamine transporter activity, which raises DA levels in the synaptic cleft, cocaine produces euphoria, enhanced energy, and alertness as acute effects (Allen et al., 2023). Seizures, status epilepticus, headaches, and strokes can result from an overdose (Poireau et al., 2022). Long-term consequences include drug craving, anxiety, depression, and hypersomnolence, which may be brought on by disruptions in the neurotransmitter system (Roque Bravo et al., 2022). Cocaine use includes neuropsychiatric aftereffects, such as cognitive impairments in executive function, memory, and decision-making, as well as effects on the respiratory, cardiovascular, and thermoregulatory systems (Jones & Graff-Radford, 2021). Cocaine misuse can cause neurological manifestations such as Tourette-like symptoms, dystonic reactions, persistent dyskinetic movements, choreic motions, and aberrant movement (Bravo et al., 2022). Additionally, some people have cocaine-induced exciting delirium, a condition that is similar to neuroleptic-induced neuroleptic malignant syndrome and is characterised by altered consciousness, autonomic dysfunction, hyperthermia, and extrapyramidal symptoms (Wilson & Vilke, 2021).

The postmortem brains of cocaine users exhibit abnormal patterns of gene and protein expression, including increased DAT levels and DA absorption in the ventral striatum (Alonso et al., 2021). Patients with cocaine-induced delirium, on the other hand, do not exhibit comparable elevations in DAT expression, indicating a subset showing distinct reactions to drug usage (de Alencar et al., 2023). The substantia nigra, caudate, putamen, and nucleus accumbens of cocaine users do not appear to have altered expression of D1 or D2 receptor genes (Clare et al., 2021). However, cocaine overdose sufferers had higher levels of D3 receptor mRNA expression in the nucleus accumbens. The striatum and nucleus accumbens also have higher concentrations of serotonin transporters (Occhipinti et al., 2023). There have also been reports of reduced dopamine uptake sites, mu opioid receptor binding, and enkephalin mRNA levels. Myelin-related genes, including myelin basic protein (MBP), proteolipid protein (PLP), and myelin-associated oligodendrocyte basic protein (MOBP), are found in lower concentrations in the brains of cocaine addicts, which may be a sign of neuroadaptive brain changes (Choi et al., 2021). Heat shock protein 70 (HSP70), an accurate predictor of excited delirium, was found to be higher in fatal cases of the illness. HSP70 may not be an effective brain biomarker for exciting delirium, nevertheless, as recent studies indicate that its expression was elevated in cocaine-related deaths (Fernandez-Castillo et al., 2022).

Abnormal prefrontal cortex (PFC) functions are a common cause of cognitive impairment in cocaine addicts (Capela & Carvalho, 2022). Reduced prefrontal cortical volume has been associated with crack cocaine and alcohol dependence, and this can result in cognitive deficits in frontal cortex-mediated functions. There can be functional repercussions from this decreased brain volume (Schinz et al., 2023). Additionally, research has shown that cocaine users' frontal cortex, temporal cortex, and cerebellum have reduced grey matter tissue density (Schinz et al., 2023). Cocaine addicts also exhibit decreased thickness and volume of the entire cortical structure, including paralimbic cortices such as the dorsal prefrontal cortex and insula (Korponay & Koenigs, 2021). Additionally, the orbitofrontal cortex, right insula, temporal gyrus, and bilateral caudate all exhibit smaller grey matter volumes in cocaine-dependent people.

Cocaine addiction may result in altered white matter integrity in the corpus callosum and frontal lobe (Thompson, 2024). Changes in white matter integrity in myelin degeneration-related disorders have been measured using diffusion tensor imaging (DTI). DTI has been used to detect a variety of tissue abnormalities; however, in order to confirm the connection between DTI and microscopic tissue changes, it must be connected to histological changes. The researchers emphasized the significance of comprehending the integrity of the white matter in the brain (Suchting et al., 2021). Bell et al. found significant structural changes in the white matter of cocaine users, including abnormalities in the bilateral superior corona radiata, right superior longitudinal fasciculus, left anterior callosal fibres, left genu of the corpus callosum, and right callosal fibres (Tondo et al., 2021). According to

Lim et al., these anomalies might result from recurrent vasoconstriction and brain hypoperfusion. They might also be brought on by the harmful side effects of long-term drug usage, like elevated cellular calcium and endoplasmic reticulum stress (Yang et al., 2025).

1.5.3 Amphetamine

Amphetamines are psychoactive drugs with stimulant, euphoric, anorectic, and hallucinogenic effects (Capela & Costa, 2023). They significantly alter monoaminergic neurotransmission and are quickly absorbed by the brain. They decrease reuptake and cause reverse transport by acting as competing substrates at noradrenaline, dopamine, and serotonin membrane transporters. To make DA, 5-HT, and NE accessible for reverse transport, amphetamines also encourage their release from storage vesicles (Risan & Ali, 2024). Whereas MDMA is typically consumed orally, methamphetamine can be taken orally, intravenously, nasally, or by smoking.

1.5.4 Methamphetamine

In 2009, estimates of the annual prevalence of methamphetamine misuse ranged from 0.3 to 1.3%, making it a global public health concern. Its lengthy duration of effect, low production cost, and long half-life all contribute to its misuse profile. After acute use, users report feeling euphoric, more productive, hypersexual, less anxious, and more energized (Li et al., 2022). Long-term adverse effects include brain damage, seizures, anxiety, sleeplessness, and paranoia. Methamphetamine abusers also have cognitive deficits that are reliant on frontostriatal and limbic circuitry (Edinoff et al., 2022). Cognitive dysfunctions, such as memory issues and self-reported deficiencies in daily functioning, are linked to methamphetamine dependency (Hayley et al., 2023). Methamphetamine causes a significant release of dopamine (DA) in the synaptic cleft, which is located in parts of the brain that receive dopaminergic projections from the midbrain (Støier et al., 2023). Postmortem studies have shown that chronic users have reduced amounts of DA, TH, and DAT in their caudate. Methamphetamine addicts' brains exhibit reduced DAT and VMAT-2 binding, according to neuroimaging studies (Kohno et al., 2022). After extended abstinence, DAT binding levels increase. The frontal cortex and anterior cingulate cortex exhibit hypometabolism, among other notable changes in brain metabolism, in methamphetamine abusers (Sanjari Moghaddam et al., 2021). Additionally, they show structural abnormalities such as decreased intracranial and hippocampal volume, altered white matter tissue integrity, and loss of cortical grey volume in the temporal and insular cortices (Jia et al., 2022). Methamphetamine misuse may raise the chance of developing Parkinson's disease (PD), according to studies. According to a recent epidemiological study, methamphetamine addicts had a 75% higher incidence of Parkinson's disease (PD) than control people (Ramli et al., 2025). This is because different parts of the brain have activated microglia and reactive astrocytosis. According to the epidemiological data, methamphetamine users' and Parkinson's disease patients' brains exhibit increased microgliosis and a significant loss of dopaminergic markers (Peng et al., 2024).

1.5.5 MDMA (3,4-Methylenedioxymethamphetamine)

Commonly used illegally in North America, Europe, and Oceania, MDMA causes euphoria, elevated self-esteem, heightened sensory experiences, tachycardia, and hyperthermia. Long-term usage, however, may result in serotonergic dysfunctions that affect working memory, executive function, verbal fluency, and prospective memory, among other cognitive areas. Psychiatric effects such as phobic anxiety, obsessionality, somatization, paranoid ideation, disturbed appetite, and restless sleep are also experienced by heavy MDMA users (Ung et al., 2025). MDMA is a chemical that prevents monoamine oxidase B from breaking down serotonin and the serotonin transporter (SERT). Additionally, it functions as an agonist for the serotonin2A receptor (Banimostafavi et al., 2024). In PET investigations, MDMA users have decreased SERT ligand binding in many brain areas. Additionally, they exhibit lower levels of SERT protein in the frontal, temporal, occipital, and striatum cortices. The brains of active MDMA users also exhibit lower numbers of 5 HT2A receptors (Coray et al., 2025). It has been demonstrated that rats given MDMA exhibit anomalies in their serotonergic system, such as reduced amounts of 5-HT and 5-HIAA, its primary metabolite, as well as fewer 5-HT transporters. These abnormalities can continue to happen months or even years after administration of medication (van de Blaak & Dumont, 2022). The MDMA abuse causes a lasting effect on the glucose metabolism of the human brain, which leads to reduced absorption in the amygdala, hippocampus, Brodmann area II, striatum, and cingulate cortex. Moreover, MDMA users have their grey matter loss in several parts of the brain, and it may be related to the cognitive decline (Raskurazhev et al., 2025). Studies have found decreased grey matter in the medial orbitofrontal cortex and medial dorsal anterior cingulate cortex and lowered grey matter in the brainstem, cerebellum, and Brodmann area. In MDMA users, the thalamus is similarly impacted (Montgomery & Roberts, 2022). MDMA neurotoxicity is linked to 5-HT2A and dopamine D1 receptor-mediated hyperthermia as well as DA-induced oxidative stress in 5-HT terminals. Although MDMA does not have a precise harmful or lethal dose, incidences of MDMA toxicity have been linked to higher concentrations. The cause of death cannot be determined without considering the history and other findings from the autopsy (Nazari et al., 2022).

1.5.6 Heroin

Approximately 16.5 million people use opiates, such as heroin and opium, globally, with a higher incidence in Eastern Europe, North America, and Southwest and Central Asia. The most often abused opiate is heroin, which is made from opium and administered intravenously, orally, or through inhalation. The physiological and biochemical actions of the drug activate opioid receptors μ, κ, and δ (Mustafa et al., 2023). While dependence quickly develops following intravenous usage, oral delivery under medical care does not cause significant adverse consequences. Disorders of the nervous system, white matter damage to the brain, and atrophy are some of the acute and chronic side effects (Ghosh et al., 2023). Postmortem studies have shown that heroin addicts had reduced immunoreactivity and density of I2-imidazoline receptors, altered G-protein subunits, and lower levels of alpha adrenergic-2 in the caudate nucleus, hypothalamus, and frontal cortex (Parker et al., 2023). These results are consistent with opiate receptor coupling to GTP-binding proteins. Furthermore, heroin addicts exhibit lower levels of NF-L protein in the frontal brain, which may indicate axonal damage brought on by long-term opiate drug consumption (Zhang et al., 2023).

Addiction to heroin causes major changes in the brain, including changes in the morphometry and functions of the grey and white matter (Briggs & Allinson, 2021). Neuronal apoptosis and decreased dendritic spine density are two clinical alterations that might result from long-term heroin consumption. Loss of frontal volume and increased ventricular gaps are further symptoms of addiction (Schaub et al., 2022). In heroin-dependent patients, methadone maintenance therapy has been shown to reduce thalamic volume and grey matter density in certain regions (Pando-Naude et al., 2021). Spongiform leukoencephalopathy is a unique condition that has been identified as a result of the inhalation of pre-heated heroin. The precise cause of this condition is still unknown, and in certain cases, the neurological impairments linked to it progress to death and are permanent (Alshamam et al., 2021). Usually, scans of these patients show symmetrical non-enhancing hypodense areas in the cerebral and cerebellar white matter. According to T1-weighted images, MRI scans show regions of reduced signal intensity (hypointense), while T2-weighted pictures show regions of increased signal intensity (hyperintense) (Knopp et al., 2023). According to neuropathological analysis, the white matter exhibits spongiform degeneration, which is characterized by a decrease in oligodendrocytes, axons, and vacuoles protected by a network of thin myelinated fibres (Macchi et al., 2022). Significant reductions in axial diffusivity in the right frontal white matter and superior longitudinal fascicule have been observed in long-term opiate abusers, indicating that prolonged usage is linked to axonal damage (Diefenbach et al., 2022). Drug-induced oxidative stress/mitochondrial dysfunction, apoptosis, and/or changes in myelin-related gene expression can all contribute to the development of white matter lesions in heroin users (Al-Sarraj et al., 2022; Cadet et al., 2014).

1.6 Challenges Faced by an Individual with Neurogenetic Disability

The various challenges that persons with disabilities experience in today's society are a result of several social obstacles that persist. These obstacles include limited access, inadequate educational systems, and financial difficulties, all of which hurt the lives of people with disabilities. For individuals with disabilities, these factors frequently result in significant consequences, including financial struggles, physical and mental health problems, social marginalization, and the perpetuation of negative stereotypes and perceptions. Successful strategies to lessen the impact of these issues lay a strong emphasis on creating settings, especially in the business and in educational institutions, that are made to accommodate people with disabilities and anticipate and address any difficulties they may encounter (Turcotte et al., 2025). In most aspects of life, mental disability provides individuals with significant barriers to accessibility. Such challenges are evident in passing through real facilities, finding employment, accessing adequate healthcare, and receiving the appropriate type of assistance in school settings. The situation of social isolation and exclusion of people with disabilities is often exacerbated by the existence of negative or limiting perceptions towards them, which are often common in the popular media and social attitudes. People with disabilities have significantly lower employment rates, leading to increased poverty rates, financial hardships, and mental health problems (Macdonald et al., 2018).

Life with disability is quite an experience with its challenges that transcend beyond the physical world. Even while accessibility and inclusivity have advanced significantly, people with impairments still face many obstacles. Individuals with disabilities face intricate challenges that impede their full participation in societal activities. The subsequent section will investigate these challenges. The psychological effects of living with a disability are a significant factor that should be considered. People may suffer from increased anxiety, social isolation, and sadness. To improve general well-being, mental health services that are specific to the unique difficulties faced by people with disabilities must be made available (Rocha Lourenço et al., n.d.). Individuals with disabilities may face different challenges while trying to access services, depending on the kind of service they need and the sort of barrier they face. According to the World Health Organization (WHO), impairment is caused by the existence of barriers (World Health Organization, 2024). Barriers are defined as aspects of a person's surroundings that limit functioning and exacerbate the manifestation of impairment, whether they are present or absent. The factors that prevent people with health conditions from participating in many aspects of life include an inaccessible physical environment, a dearth of pertinent assistive technology (such as assistive, adaptive, and rehabilitative devices), negative societal perceptions of disability and related services, and inadequate or ineffective systems and policies (World Health Organization, 2007). According to the US Centres for Disease Control and Prevention (CDC), it is common for several barriers to arise at the same time. The following is a list of the most common barriers faced by disabled individuals (Curtis et al., 2016; Calvert, 2021):

- **Attitude:** It includes prejudice and discrimination, as well as stigma and stereotyping. The CDCP defines stereotyping as the assumption that someone with a disability must lead a poor life.
- **Communication:** Communication difficulties may arise for those with disabilities that impact their ability to communicate. They include printed communications that could be difficult for people with visual impairments to access, and the use of "technical language, lengthy sentences, and polysyllabic words", which could be difficult for people with cognitive impairments to understand.
- **Educational barriers:** Despite legislative efforts to promote inclusive education, obstacles persist in educational institutions. The academic challenges faced by individuals with disabilities are largely caused by the existence of inaccessible facilities, inadequate accommodations, and a shortage of teachers with the necessary training.
- **Physical:** Physical obstacles limit access to particular resources or places or hinder or block a person's mobility. For instance, curbs and steps that prevent a person with a mobility impairment from using a pavement or entering a building. Moreover, certain patients may find it challenging to undergo therapy due to physical obstacles, such as "mammography equipment that requires a woman with mobility impairment to stand".
- **Limited accessibility:** The accessibility of buildings, public areas, and transit networks remains a major obstacle for people with disabilities. The mobility and autonomy of people with disabilities are sometimes limited by inadequate infrastructure and the lack of places that are broadly designed.
- **Financial strains:** Additional financial obligations, such as medical bills, assistive technology, and specialized caregiving services, might arise from having a disability. The occurrence of financial stress is a severe problem that necessitates strengthening social support systems to alleviate the financial burden on individuals and their families.
- **Technological gaps:** Even if technology has the potential to empower people with disabilities, there is still a significant gap in access. The accessibility and affordability of assistive devices significantly limit the ability of people with disabilities to fully utilize the wide range of technological advancements.
- **Policy:** People frequently mention the ignorance or non-implementation of the laws and regulations that now require programs and activities for persons with disabilities as the cause of the policy barrier.
- **Programmatic:** The program aims to facilitate the provision of public health programs to individuals with various forms of impairment. Some examples provided by the CDCP are a shortage of readily available equipment and time allocated to medical inspection and surgery.
- **Social:** Social determinants of health are the conditions in which people are born, grow up, live, learn, work, and age. These aspects are linked to social obstacles and may result in the impaired performance of the disabled. The data

provided by the Centers for Disease Control and Prevention (CDCP) shows that people with disabilities are less likely to be employed. Discrimination and misconceptions about disability conditions significantly contribute to the social exclusion of people with disabilities. The stereotypes can cause people to become acquired into a cycle of exclusion and isolation, as they are unable to integrate socially, work, or have an education.

- **Employment disparities:** Individuals with disabilities can encounter significant barriers at the workplace. Discrimination, inaccessible environments, and the absence of relevant changes are known to reduce people's employment opportunities. Employers and legislators need to collaborate to overcome these limitations.
- **Healthcare disparities:** Impaired individuals may negatively affect the accessibility of quality healthcare services. Some of the barriers include the location of the medical facilities that are not accessible physically, the absence of readily available information, and the disregard of certain medical needs. Healthcare providers must bridge this gap to safeguard the extensive advantages experienced by individuals with disabilities.
- **Transportation:** The barriers to transportation hinder the independence of an individual, such as a person with sight problems who will not be able to drive, and the provision of accessible or convenient means of transport will be a major impediment. Public transport systems do not consider the needs of individuals with impairments at times. The lack of such features as wheelchair ramps and loud announcements impedes autonomous mobility.

1.7 Conclusion

In general, neurological disability, neuro-disability, and neurogenetic disability constitute a broad category of disorders that are caused by genetic, developmental, vascular, traumatized, and substance-related damage to the nervous system, which leads to severe dysfunctions and disability in everyday life. Disorders have an enormous impact on the functioning of the brain, cognition, behaviour, and overall development, diminishing personal abilities and living standards. Genetic integrity is crucial to normal neural development, as shown by neurogenetic and neurodevelopmental disorders like disability as a result of chromosome disorders, including Down Syndrome (Trisomy 21), Trisomy 18 (Edwards Syndrome), Trisomy 13 (Patau Syndrome), 45, X (Turner Syndrome), and Triploidy (69, XXX or 69, XXY). Other conditions, including acquired neurological disabilities like stroke, traumatic brain injury (TBI), chronic traumatic encephalopathy (CTE), extra-axial hematoma, traumatic subarachnoid haemorrhage, and diffuse axonal injury (DAI), again demonstrate how the brain is prone to damage and the effects it has in the long run. In addition, substance or drug abuse neuro-disability, e.g. marijuana (cannabis) use disorder, cocaine use disorder, amphetamine analogue use disorders, methamphetamine use disorder, MDMA use disorder, and heroin use, points to the effect of neurotoxic exposures, which are preventable but mortal. Each of these conditions, in its turn, presupposes early diagnosis, interdisciplinary treatment, and long-term rehabilitation. To offer holistic and dignified care, it is also significant to address the societal dimension of the disability of the affected individuals and their families, such as stigma and inaccessibility of non-discriminatory education, health, and employment.

References

Abad, L. H., Lessa, R. T., de Mesquita, F. B., Silva, V. L., Cesar, M. R., Ferreira, T. B., da Costa, A. J., Buzan, T. N., & Santo Mendes, N. B. (2022). Chronic traumatic encephalopathy in sports practice: A literature review. *Arquivos Brasileiros de Neurocirurgia: Brazilian Neurosurgery, 41*(04), e362–e367.

Akira, M., Yuichi, T., Tomotaka, U., Takaaki, K., Kenichi, M., & Chimi, M. (2022). The outcome of neurorehabilitation efficacy and management of traumatic brain injury. *Frontiers in Human Neuroscience, 16*, 870190.

Allen, M. I., Duke, A. N., Nader, S. H., Adler-Neal, A., Solingapuram Sai, K. K., Reboussin, B. A., Gage, H. D., Voll, R. J., Mintz, A., Goodman, M. M., & Nader, M. A. (2023). PET imaging of dopamine transporters and D2/D3 receptors in female monkeys: Effects of chronic cocaine self-administration. *Neuropsychopharmacology, 48*(10), 1436–1445.

Allick, A., Park, G., Kim, K., Vintimilla, M., Rathod, K., Lebo, R., Nanavati, J., & Hammond, C. J. (2021). Age-and sex-related cortical gray matter volume differences in adolescent cannabis users: A systematic review and meta-analysis of voxel-based morphometry studies. *Frontiers in Psychiatry, 12*, 745193.

Alonso, I. P., Pino, J. A., Kortagere, S., Torres, G. E., & España, R. A. (2021). Dopamine transporter function fluctuates across sleep/wake state: Potential impact for addiction. *Neuropsychopharmacology, 46*(4), 699–708.

Al-Sarraj, S., Troakes, C., & Rutty, G. N. (2022). Axonal injury is detected by βAPP immunohistochemistry in rapid death from head injury following road traffic collision. *International Journal of Legal Medicine, 136*(5), 1321–1339.

Alshamam, M. S., Sumbly, V., Nso, N., Saliaj, M., & Gurung, D. O. (2021). Heroin-induced leukoencephalopathy. *Cureus, 13*(2).

Anaby, D., Hand, C., Bradley, L., DiRezze, B., Forhan, M., DiGiacomo, A., & Law, M. (2013). The effect of the environment on participation of children and youth with disabilities: A scoping review. *Disability and rehabilitation., 35*(19), 1589–1598.

Anacleto, P. Z., Sousa, V. A., & Joviano-Santos, J. V. (2024). Trisomy 13, home health-care and multidisciplinary approach: Case report. *Revista Paulista de Pediatria, 43*, e2024018.

Armour, E., MacPherson, M. J., Mack, C., & van Manen, M. (2025). Parental perspectives of trisomy 18: Common threads of a life-limiting diagnosis. *Medical Humanities*.

Banimostafavi, E. S., Taheri, A., Malakian, A., & Zakariaei, Z. (2024). Massive intracerebral hemorrhage following methamphetamine poisoning in a man: A case report. *International Journal of Surgery Case Reports, 117*, 109489.

Baricich, A., & Carda, S. (2018). Chronic stroke: An oxymoron or a challenge for rehabilitation? *Functional Neurology, 33*(3), 123–124.

Baricich, A., Spitoni, G. F., & Morone, G. (2022). Long term disability in neurological disease: A rehabilitation perspective. *Frontiers in Neurology, 13*, 964664.

Batalla, A., Bhattacharyya, S., Yuecel, M., Fusar-Poli, P., Crippa, J. A., Nogue, S., Torrens, M., Pujol, J., Farre, M., & Martin-Santos, R. (2013). Structural and functional imaging studies in chronic cannabis users: A systematic review of adolescent and adult findings. *PLoS One, 8*(2), e55821.

Bejjani, B. A., & Shaffer, L. G. (2008). Clinical utility of contemporary molecular cytogenetics. *Annual Review of Genomics and Human Genetics, 9*(1), 71–86.

Brassard, S. L., Dosanjh, J., Cooper, J., Weber, J., Zald, D., MacKillop, J., & Balodis, I. M. (2025). A Behavioural and neurobiological assessment of effort-based decision-making in cannabis use disorder: An initial/preliminary investigation. *Cognitive, Affective, & Behavioral Neuroscience*, 1–9.

Bravo, R. R., Faria, A. C., Brito-da-Costa, A. M., Carmo, H., Mladěnka, P., da Silva, D. D., & Remião, F. (2022). Cocaine: An updated overview on chemistry, detection, biokinetics, and pharmacotoxicological aspects including abuse pattern. *Toxins, 14*(4), 278.

Briggs, M., & Allinson, K. (2021). Fatal toxic leukoencephalopathy and bilateral basal ganglia necrosis associated with inhaled heroin and cocaine use: A case report. *Clinical Neuropathology, 40*(6), 361.

Bull, M. J. (2011). Committee on genetics. Health supervision for children with Down syndrome. *Pediatrics, 128*(2), 393–406.

Cadet, J. L., Bisagno, V., & Milroy, C. M. (2014). Neuropathology of substance use disorders. *Acta Neuropathologica, 127*(1), 91–107.

Calvert, S. (2021). Challenges for people with disabilities. *Ballard Brief, 2021*(3), 6.

Capela, J. P., & Carvalho, F. D. (2022). A review on the mitochondrial toxicity of "ecstasy" (3, 4-methylenedioxymethamphetamine, MDMA). *Current Research in Toxicology, 3*, 100075.

Capela, J. P., & Costa, V. M. (2023). Pharmacology and toxicology of amphetamine-type stimulants. *Future Pharmacology, 3*(2), 515–516.

Casanova, E. L., Gerstner, Z., Sharp, J. L., Casanova, M. F., & Feltus, F. A. (2018). Widespread genotype-phenotype correlations in intellectual disability. *Frontiers in Psychiatry, 9*, 535.

Chen, W. J., Cheng, X., Fu, Y., Zhao, M., McGinley, J., Westenberger, A., & Xiong, Z. Q. (2020). Rethinking monogenic neurological diseases. *BMJ*, 371.

Choi, M. R., Jin, Y. B., Kim, H. N., Chai, Y. G., Im, C. N., Lee, S. R., & Kim, D. J. (2021). Gene expression in the striatum of cynomolgus monkeys after chronic administration of cocaine and heroin. *Basic & Clinical Pharmacology & Toxicology, 128*(5), 686–698.

Clare, K., Pan, C., Kim, G., Park, K., Zhao, J., Volkow, N. D., Lin, Z., & Du, C. (2021). Cocaine reduces the neuronal population while upregulating dopamine D2-receptor-expressing neurons in brain reward regions: Sex-effects. *Frontiers in Pharmacology, 12*, 624127.

Coray, R. C., Beliveau, V., Zimmermann, J., Preller, K. H., Wunderli, M., Baumgartner, M. R., Seifritz, E., Stock, A. K., Beste, C., Cole, D. M., & Quednow, B. B. (2025). Memory deficits of MDMA users are linked to cortical thinning related to 5-HT receptor densities. *Brain*, awaf391.

Curtis, K. M., Tepper, N. K., Jatlaoui, T. C., & Whiteman, M. K. (2016). Removing medical barriers to contraception—evidence-based recommendations from the Centers for Disease Control and Prevention, 2016. *Contraception, 94*(6), 579–581.

Daglish, M. R., & Nutt, D. J. (2003). Brain imaging studies in human addicts. *European Neuropsychopharmacology, 13*(6), 453–458.

de Alencar, J. C., Garcez, F. B., Pinto, A. A., Silva, L. O., Soler, L. D., Fernandez, S. S., Van Vaisberg, V., Gomez Gomez, L. M., Ribeiro, S. M., Avelino-Silva, T. J., & Souza, H. P. (2023). Brain injury biomarkers do not predict delirium in acutely ill older patients: A prospective cohort study. *Scientific Reports, 13*(1), 4964.

Devinsky, O., Coller, J., Ahrens-Nicklas, R., Liu, X. S., Ahituv, N., Davidson, B. L., Bishop, K. M., Weiss, Y., & Mingorance, A. (2025). Gene therapies for neurogenetic disorders. *Trends in Molecular Medicine*.

Diefenbach, C., Lorenz, C., & Weidauer, S. (2022). Toxic spongiform leukoencephalopathy after intravenous heroin abuse: Unusual but important differential diagnosis of acute impairment of consciousness. *Fortschritte Der Neurologie-Psychiatrie, 90*(11), 523–528.

Dobkin, B. H. (2005). Rehabilitation after stroke. *New England Journal of Medicine, 352*(16), 1677–1684.

Dong, W., Liu, Y., Bai, R., Zhang, L., & Zhou, M. (2025). The prevalence and associated disability burden of mental disorders in children and adolescents in China: A systematic analysis of data from the global burden of disease study. *The Lancet Regional Health-Western Pacific*, 55.

Dos Santos, J. C., Mano, G. B., da Cunha Barreto-Vianna, A. R., Garcia, T. F., de Vasconcelos, A. V., Sá, C. S., de Souza Santana, S. L., Farias, A. G., Seimaru, B., Lima, M. P., & Goes, J. V. (2024). The molecular impact of glucosylceramidase beta 1 (Gba1) in Parkinson's disease: A new genetic state of the art. *Molecular Neurobiology, 61*(9), 6754–6770.

Du, X., Gao, X., Liu, X., Shen, L., Wang, K., Fan, Y., Sun, Y., Luo, X., Liu, H., Wang, L., & Wang, Y. (2018). Genetic diagnostic evaluation of trio-based whole exome sequencing among children with diagnosed or suspected autism spectrum disorder. *Frontiers in Genetics, 9*, 594.

Edinoff, A. N., Kaufman, S. E., Green, K. M., Provenzano, D. A., Lawson, J., Cornett, E. M., Murnane, K. S., Kaye, A. M., & Kaye, A. D. (2022). Methamphetamine use: A narrative review of adverse effects and related toxicities. *Health Psychology Research, 10*(3), 38161.

Elipe-Lorenzo, P., Diez-Fernández, P., Ruibal-Lista, B., & López-García, S. (2025). Barriers faced by people with disabilities in mainstream sports: A systematic review. *Frontiers in Sports and Active Living, 7*, 1520962.

ElSohly, M. A., & Slade, D. (2005). Chemical constituents of marijuana: The complex mixture of natural cannabinoids. *Life Sciences, 78*(5), 539–548.

Emanuel, B. S., & Saitta, S. C. (2007). From microscopes to microarrays: Dissecting recurrent chromosomal rearrangements. *Nature Reviews Genetics, 8*(11), 869–883.

Emanuel, B. S., & Shaikh, T. H. (2001). Segmental duplications: An 'expanding' role in genomic instability and disease. *Nature Reviews Genetics, 2*(10), 791–800.

Evanski, J. M., Zundel, C. G., Baglot, S. L., Desai, S., Gowatch, L. C., Ely, S. L., Sadik, N., Lundahl, L. H., Hill, M. N., & Marusak, H. A. (2024). The first "hit" to the endocannabinoid system? Associations between prenatal cannabis exposure and frontolimbic white matter pathways in children. *Biological Psychiatry Global Open Science, 4*(1), 11–18.

Fernandes, A., Mourato, A., Xavier, M., Andrade, D., Fernandes, C., & Palha, M. (2001). Characterisation of the somatic evolution of Portuguese children with trisomy 21-preliminary results. *Down Syndrome Research and Practice, 6*(3), 134–138.

Fernandez-Castillo, N., Cabana-Dominguez, J., Corominas, R., & Cormand, B. (2022). Molecular genetics of cocaine use disorders in humans. *Molecular Psychiatry, 27*(1), 624–639.

Fleisher-Berkovich, S., Battaglia, V., Baratta, F., Brusa, P., Ventura, Y., Sharon, N., Dahan, A., Collino, M., & Ben-Shabat, S. (2023). An emerging strategy for neuroinflammation treatment: Combined Cannabidiol and angiotensin receptor blockers treatments effectively inhibit glial nitric oxide release. *International Journal of Molecular Sciences, 24*(22), 16254.

Fong, C. T., & Brodeur, G. M. (1987). Down's syndrome and leukemia: Epidemiology, genetics, cytogenetics and mechanisms of leukemogenesis. *Cancer Genetics and Cytogenetics, 28*(1), 55–76.

Fu, J. M., Satterstrom, F. K., Peng, M., Brand, H., Collins, R. L., Dong, S., Wamsley, B., Klei, L., Wang, L., Hao, S. P., & Stevens, C. R. (2022). Rare coding variation provides insight into the genetic architecture and phenotypic context of autism. *Nature Genetics, 54*(9), 1320–1331.

Ghosh, A., Shaktan, A., Nehra, R., Basu, D., Verma, A., Rana, D. K., Modi, M., & Ahuja, C. K. (2023). Heroin use and neuropsychological impairments: Comparison of intravenous and inhalational use. *Psychopharmacology, 240*(4), 909–920.

Giess, D., Erdos, J., & Wild, C. (2024). An updated systematic review on spinal muscular atrophy patients treated with nusinersen, onasemnogene abeparvovec (at least 24 months), risdiplam (at least 12 months) or combination therapies. *European Journal of Paediatric Neurology, 51*, 84–92.

Gilissen, C., Hehir-Kwa, J. Y., Thung, D. T., Van De Vorst, M., Van Bon, B. W., Willemsen, M. H., Kwint, M., Janssen, I. M., Hoischen, A., Schenck, A., & Leach, R. (2014). Genome sequencing identifies major causes of severe intellectual disability. *Nature, 511*(7509), 344–347.

Glass, M., Faull, R. L., & Dragunow, M. (1997). Cannabinoid receptors in the human brain: A detailed anatomical and quantitative autoradiographic study in the fetal, neonatal and adult human brain. *Neuroscience, 77*(2), 299–318.

Gok, H., Celik, S. E., Yangi, K., Yavuz, A. Y., Percinoglu, G., Unlu, N. U., & Goksu, K. (2023). Management of epidural hematomas in pediatric and adult population: A hospital-based retrospective study. *World Neurosurgery, 177*, e686–e692.

Gold, W. A., Percy, A. K., Neul, J. L., Cobb, S. R., Pozzo-Miller, L., Issar, J. K., Ben-Zeev, B., Vignoli, A., & Kaufmann, W. E. (2024). Rett syndrome. *Nature Reviews Disease Primers, 10*(1), 84.

Gómez Sánchez, L. E., Schalock, R. L., & Verdugo, M. Á. (2021). A new paradigm in the field of intellectual and developmental disabilities: Characteristics and evaluation. *Psicothema.*

Gravholt, C. H., Viuff, M., Just, J., Sandahl, K., Brun, S., van der Velden, J., Andersen, N. H., & Skakkebaek, A. (2023). The changing face of Turner syndrome. *Endocrine Reviews, 44*(1), 33–69.

Greenberg, M. S. (2006). *Handbook of neurosurgery* (7th ed., pp. 289–365). Mark S. Greenberg Thieme.

Gunatilake, N., Boyes, A., Elwyn, R., Brunet, A., Hermens, D. F., & Driver, C. (2025). Adolescent brain development following early life stress: A systematic review of white matter alterations from diffusion tensor imaging studies. *Adolescent Research Review*, 1–48.

Guo, H., Wang, T., Wu, H., Long, M., Coe, B. P., Li, H., Xun, G., Ou, J., Chen, B., Duan, G., & Bai, T. (2018). Inherited and multiple de novo mutations in autism/developmental delay risk genes suggest a multifactorial model. *Molecular Autism, 9*(1), 64.

Haldeman-Englert, C. R., Saitta, S. C., & Zackai, E. H. (2018). Chromosome disorders. In *Avery's diseases of the newborn* (pp. 211–223). Elsevier.

Hall, W., & Degenhardt, L. (2009). Adverse health effects of non-medical cannabis use. *The Lancet, 374*(9698), 1383–1391.

Hardisty, E. E., & Vora, N. L. (2014). Advances in genetic prenatal diagnosis and screening. *Current Opinion in Pediatrics, 26*(6), 634–638.

Hayley, A. C., Shiferaw, B., Rositano, J., & Downey, L. A. (2023). Acute neurocognitive and subjective effects of oral methamphetamine with low doses of alcohol: A randomised controlled trial. *Journal of Psychopharmacology, 37*(9), 928–936.

Head, E., Lott, I. T., Wilcock, D. M., & Lemere, C. A. (2016). Aging in Down syndrome and the development of Alzheimer's disease neuropathology. *Current Alzheimer Research, 13*(1), 18–29.

Hook, E. (1992). Ultrasound and fetal chromosome abnormalities. *The Lancet, 340*(8827), 1109.

Hook, E. B., & Cross, P. K. (1982). Paternal age and Down's syndrome genotypes diagnosed prenatally: No association in New York state data. *Human Genetics, 62*(2), 167–174.

Hormozdiari, F., Penn, O., Borenstein, E., & Eichler, E. E. (2015). The discovery of integrated gene networks for autism and related disorders. *Genome Research, 25*(1), 142–154.

Huang, Y., Zhao, Y., Ren, Y., Yi, Y., Li, X., Gao, Z., Zhan, X., Yu, J., Wang, D., Liang, S., & Wu, L. (2019). Identifying genomic variations in monozygotic twins discordant for autism spectrum disorder using whole-genome sequencing. *Molecular Therapy Nucleic Acids, 14*, 204–211.

Iossifov, I., Levy, D., Allen, J., Ye, K., Ronemus, M., Lee, Y. H., Yamrom, B., & Wigler, M. (2015). Low load for disruptive mutations in autism genes and their biased transmission. *National Academy of Sciences of the United States of America, 112*(41), E5600–E5607.

Iyer, J., Singh, M. D., Jensen, M., Patel, P., Pizzo, L., Huber, E., Koerselman, H., Weiner, A. T., Lepanto, P., Vadodaria, K., & Kubina, A. (2018). Pervasive genetic interactions modulate neurodevelopmental defects of the autism-associated 16p11. 2 deletion in Drosophila melanogaster. *Nature Communications, 9*(1), 2548.

Jia, X., Wang, J., Jiang, W., Kong, Z., Deng, H., Lai, W., Ye, C., Guan, F., Li, P., Zhao, M., & Yang, M. (2022). Common gray matter loss in the frontal cortex in patients with methamphetamine-associated psychosis and schizophrenia. *NeuroImage: Clinical, 36*, 103259.

Jones, D. T., & Graff-Radford, J. (2021). Executive dysfunction and the prefrontal cortex. *CONTINUUM: Lifelong Learning in Neurology, 27*(6), 1586–1601.

Karaca, E., Posey, J. E., Coban Akdemir, Z., Pehlivan, D., Harel, T., Jhangiani, S. N., Bayram, Y., Song, X., Bahrambeigi, V., Yuregir, O. O., & Bozdogan, S. (2018). Phenotypic expansion illuminates multilocus pathogenic variation. *Genetics in Medicine, 20*(12), 1528–1537.

Knopp, B. W., Weiss, H. Z., Retrouvey, M., Luck, G., & Weiss, H. (2023). Inhalational heroin use and leukoencephalopathy: A case report. *Cureus, 15*(7).

Kohno, M., Dennis, L. E., McCready, H., & Hoffman, W. F. (2022). Dopamine dysfunction in stimulant use disorders: Mechanistic comparisons and implications for treatment. *Molecular Psychiatry, 27*(1), 220–229.

Korponay, C., & Koenigs, M. (2021). Gray matter correlates of impulsivity in psychopathy and in the general population differ by kind, not by degree: A comparison of systematic reviews. *Social Cognitive and Affective Neuroscience, 16*(7), 683–695.

Kosho, T., & Carey, J. C. (2016). Does medical intervention affect outcome in infants with trisomy 18 or trisomy 13? *American Journal of Medical Genetics Part A, 170*(4), 847–849.

Kurki, M. I., Saarentaus, E., Pietiläinen, O., Gormley, P., Lal, D., Kerminen, S., Torniainen-Holm, M., Hämäläinen, E., Rahikkala, E., Keski-Filppula, R., & Rauhala, M. (2019). Contribution of rare and common variants to intellectual disability in a sub-isolate of northern Finland. *Nature Communications, 10*(1), 410.

Kwakkel, G., Kollen, B., & Lindeman, E. (2004). Understanding the pattern of functional recovery after stroke: Facts and theories. *Restorative Neurology and Neuroscience, 22*(3–5), 281–299.

Langhorne, P., Bernhardt, J., & Kwakkel, G. (2011). Stroke rehabilitation. *The Lancet, 377*(9778), 1693–1702.

Lee, W. S., Stephenson, S. E., Howell, K. B., Pope, K., Gillies, G., Wray, A., Maixner, W., Mandelstam, S. A., Berkovic, S. F., Scheffer, I. E., & MacGregor, D. (2019). Second-hit DEPDC5 mutation is limited to dysmorphic neurons in cortical dysplasia type IIA. *Annals of Clinical and Translational Neurology, 6*(7), 1338–1344.

Li, M. H., Zhang, M., Yuan, T. F., & Rao, L. L. (2022). Impaired social decision-making in males with methamphetamine use disorder. *Addiction Biology, 27*(5), e13204.

Liu, P., Meng, L., Normand, E. A., Xia, F., Song, X., Ghazi, A., Rosenfeld, J., Magoulas, P. L., Braxton, A., Ward, P., & Dai, H. (2019). Reanalysis of clinical exome sequencing data. *New England Journal of Medicine, 380*(25), 2478–2480.

Lovras, M., Rana, A., Rani, S., Chauhan, A., Sridhar, S. B., Rajput, S., Malviya, R., & Wadhwa, T. (2025). Advancement in gene therapy for the treatment of Parkinson's disease: A comprehensive review. *Mini-reviews in Medicinal Chemistry.*

Macchi, Z. A., Carlisle, T. C., & Filley, C. M. (2022). Prognosis in substance abuse-related acute toxic leukoencephalopathy: A scoping review. *Journal of the Neurological Sciences, 442*, 120420.

Macdonald, S. J., Deacon, L., Nixon, J., Akintola, A., Gillingham, A., Kent, J., Ellis, G., Mathews, D., Ismail, A., Sullivan, S., & Dore, S. (2018). 'The invisible enemy': Disability, loneliness and isolation. *Disability & Society, 33*(7), 1138–1159.

MacDonald, M., Hassold, T., Harvey, J., Wang, L. H., Morton, N. E., & Jacobs, P. (1994). The origin of 47, XXY and 47, XXX aneuploidy: Heterogeneous mechanisms and role of aberrant recombination. *Human Molecular Genetics, 3*(8), 1365–1371.

Madhok, D. Y., Rodriguez, R. M., Barber, J., Temkin, N. R., Markowitz, A. J., Kreitzer, N., Manley, G. T., Badjatia, N., Duhaime, A. C., Feeser, V. R., & Ferguson, A. R. (2022). Outcomes in patients with mild traumatic brain injury without acute intracranial traumatic injury. *JAMA Network Open, 5*(8), e2223245.

Maldonado, R., Berrendero, F., Ozaita, A., & Robledo, P. (2011). Neurochemical basis of cannabis addiction. *Neuroscience, 181*, 1–7.

Marinho, P. (n.d.) *Differential effects of cannabis vapour constituents on brain connectivity: Exploring the long-term effects of adolescent exposure* (Master's thesis, The University of Western Ontario (Canada)).

Masel BE, DeWitt DS, Levin H, Shum D, Chan R. Traumatic brain injury disease: Long-term consequences of traumatic brain injury. Understanding traumatic brain injury: Current research and future directions. 2014;28.

Mesfin, F. B., Gupta, N., Shapshak, A. H., & Margetis, K. Diffuse axonal injury. InStatPearls [Internet] 2025 Jul 7. StatPearls Publishing.

Montgomery, C., & Roberts, C. A. (2022). Neurological and cognitive alterations induced by MDMA in humans. *Experimental Neurology, 347*, 113888.

Moore, T. H., Zammit, S., Lingford-Hughes, A., Barnes, T. R., Jones, P. B., Burke, M., & Lewis, G. (2007). Cannabis use and risk of psychotic or affective mental health outcomes: A systematic review. *The Lancet, 370*(9584), 319–328.

Morris, C., Janssens, A., Tomlinson, R., Williams, J., & Logan, S. (2013). Towards a definition of neurodisability: A Delphi survey. *Developmental Medicine & Child Neurology, 55*(12), 1103–1108.

Mosleh, R., Labella Álvarez, F., Bouthour, W., Saindane, A. M., Dattilo, M., Bruce, B. B., Newman, N. J., & Biousse, V. (2024). Glaucoma as a cause of optic nerve abnormalities on magnetic resonance imaging. *Eye, 38*(9), 1626–1632.

Munro, S., Thomas, K. L., & Abu-Shaar, M. (1993). Molecular characterization of a peripheral receptor for cannabinoids. *Nature, 365*(6441), 61–65.

Murray, J., Ashworth, R., Forster, A., & Young, J. (2003). Developing a primary care-based stroke service: A review of the qualitative literature. *The British Journal of General Practice, 53*(487), 137.

Mustafa, S., Bajic, J. E., Barry, B., Evans, S., Siemens, K. R., Hutchinson, M. R., & Grace, P. M. (2023). One immune system plays many parts: The dynamic role of the immune system in chronic pain and opioid pharmacology. *Neuropharmacology, 228*, 109459.

Nazari, Z., Bahrehbar, K., & Golalipour, M. J. (2022). Effect of MDMA exposure during pregnancy on cell apoptosis, astroglia, and microglia activity in rat offspring striatum. *Iranian Journal of Basic Medical Sciences, 25*(9), 1091.

Niemi, M. E., Martin, H. C., Rice, D. L., Gallone, G., Gordon, S., Kelemen, M., McAloney, K., McRae, J., Radford, E. J., Yu, S., & Gecz, J. (2018). Common genetic variants contribute to risk of rare severe neurodevelopmental disorders. *Nature, 562*(7726), 268–271.

Nussbaum, R. L., McInnes, R. R., & Willard, H. F. (2007). *Thompson & Thompson genetics in medicine*. Saunders.

Occhipinti, C., La Russa, R., Iacoponi, N., Lazzari, J., Costantino, A., Di Fazio, N., Del Duca, F., Maiese, A., & Fineschi, V. (2023). miRNAs and substances abuse: Clinical and forensic pathological implications: A systematic review. *International Journal of Molecular Sciences, 24*(23), 17122.

Pando-Naude, V., Toxto, S., Fernandez-Lozano, S., Parsons, C. E., Alcauter, S., & Garza-Villarreal, E. A. (2021). Gray and white matter morphology in substance use disorders: A neuroimaging systematic review and meta-analysis. *Translational Psychiatry, 11*(1), 29.

Papini, M. G., Avila, A. N., Fitzgerald, M., & Hellewell, S. C. (2025). Evidence for altered white matter organization after mild traumatic brain injury: A scoping review on the use of diffusion magnetic resonance imaging and blood-based biomarkers to investigate acute pathology and relationship to persistent post-concussion symptoms. *Journal of Neurotrauma, 42*(7–8), 640–667.

Parker, C. A., Nutt, D. J., & Tyacke, R. J. (2023). Imidazoline-I2 PET tracers in neuroimaging. *International Journal of Molecular Sciences, 24*(12), 9787.

Pasin, T., & Dogruoz Karatekin, B. (2024). Determinants of social participation in people with disability. *PLoS One, 19*(5), e0303911.

Peng, Y., Yang, G., Wang, S., Lin, W., Zhu, L., Dong, W., Shen, B., Nie, Q., Hong, S., & Li, L. (2024). Triggering receptor expressed on myeloid cells 2 deficiency exacerbates methamphetamine-induced activation of microglia and neuroinflammation. *International Journal of Toxicology, 43*(2), 165–176.

Piedade, S. R., Hutchinson, M. R., Ferreira, D. M., Cristante, A. F., & Maffulli, N. (2021). The management of concussion in sport is not standardized. A systematic review. *Journal of Safety Research, 76*, 262–268.

Pizzo, L., Jensen, M., Polyak, A., Rosenfeld, J. A., Mannik, K., Krishnan, A., McCready, E., Pichon, O., Le Caignec, C., Van Dijck, A., & Pope, K. (2019). Rare variants in the genetic background modulate cognitive and developmental phenotypes in individuals carrying disease-associated variants. *Genetics in Medicine, 21*(4), 816–825.

Poireau, M., Milpied, T., Maillard, A., Delmaire, C., Volle, E., Bellivier, F., Icick, R., Azuar, J., Marie-Claire, C., Bloch, V., & Vorspan, F. (2022). Biomarkers of relapse in cocaine use disorder: A narrative review. *Brain Sciences, 12*(8), 1013.

Posey, J. E., Harel, T., Liu, P., Rosenfeld, J. A., James, R. A., Coban Akdemir, Z. H., Walkiewicz, M., Bi, W., Xiao, R., Ding, Y., & Xia, F. (2017). Resolution of disease phenotypes resulting from multilocus genomic variation. *New England Journal of Medicine, 376*(1), 21–31.

Powers, W. J., Rabinstein, A. A., Ackerson, T., Adeoye, O. M., Bambakidis, N. C., Becker, K., Biller, J., Brown, M., Demaerschalk, B. M., Hoh, B., & Jauch, E. C. (2018). Guidelines for the early management of patients with acute ischemic stroke: A guideline for healthcare professionals from the American Heart Association/American Stroke Association. *Stroke, 49*(3), e46–e99.

Pueschel, S. M., Bernier, J. C., & Pezzullo, J. C. (1991). Behavioural observations in children with Down's syndrome. *Journal of Intellectual Disability Research, 35*(6), 502–511.

Radley, J. A., O'Sullivan, R. B., Turton, S. E., Cox, H., Vogt, J., Morton, J., Jones, E., Smithson, S., Lachlan, K., Rankin, J., & Clayton-Smith, J. (2019). Deep phenotyping of 14 new patients with IQSEC2 variants, including monozygotic twins of discordant phenotype. *Clinical Genetics, 95*(4), 496–506.

Rajput, S., Malviya, R., Sridhar, S. B., Wadhwa, T., Shareef, J., & Rana, A. (2025). Advanced miRNA-nanoparticle strategies for brain cancer treatment: Bypass the blood-brain barrier for efficient treatment. *Nano-Structures & Nano-Objects, 43*, 101516.

Ramli, F. F., Rejeki, P. S., Abdullayeva, G., & Halim, S. (2025). A mechanistic review on toxicity effects of methamphetamine. *International Journal of Medical Sciences, 22*(3), 482.

Rana, A., Malviya, R., Rajput, S., Sridhar, S. B., & Wadhwa, T. (2025). Trends in nanoparticle-based strategies for the management of neuroinflammation. *CNS and Neurological Disorders: Drug Targets*.

Rana, A., Mittal, A., Vashist, C., Sharma, S., Rajput, S., Sridhar, S. B., & Malviya, R. (2025). Nanoparticle-based approaches for glioblastoma treatment: Advances and future prospects. *CNS and Neurological Disorders-Drug Targets*.

Raskurazhev, A. A., Tanashyan, M. M., Morozova, S. N., Kuznetsova, P. I., Annushkin, V. A., Mazur, A. S., Panina, A. A., Spryshkov, N. E., & Piradov, M. A. (2025). Pharmacological functional MRI technology: Potential for use in neurology. *Annals of Clinical and Experimental Neurology, 19*(1), 68–76.

Reichard, J., & Zimmer-Bensch, G. (2021). The epigenome in neurodevelopmental disorders. *Frontiers in Neuroscience, 15*, 776809.

Reinson, K., Õiglane-Shlik, E., Talvik, I., Vaher, U., Õunapuu, A., Ennok, M., Teek, R., Pajusalu, S., Murumets, Ü., Tomberg, T., & Puusepp, S. (2016). Biallelic CACNA1A mutations cause early onset epileptic encephalopathy with progressive cerebral, cerebellar, and optic nerve atrophy. *American Journal of Medical Genetics Part A, 170*(8), 2173–2176.

Ribierre, T., Deleuze, C., Bacq, A., Baldassari, S., Marsan, E., Chipaux, M., Muraca, G., Roussel, D., Navarro, V., Leguern, E., & Miles, R. (2018). Second-hit mosaic mutation in mTORC1 repressor DEPDC5 causes focal cortical dysplasia–associated epilepsy. *The Journal of Clinical Investigation, 128*(6), 2452–2458.

Risan, T. Z., & Ali, B. M. Study the effect of amphetamine on neurotransmitter factors in abusers individuals. In AIP conference proceedings 2024 Mar 8 (Vol. 3092, no. 1, p. 030005). AIP Publishing LLC.

Rocha Lourenço, F., Oliveira, R., & Tymoshchuk, O. (n.d.). Challenges and gaps in promoting inclusive spaces: A study based on interviews. In *International conference on human-computer interaction* (pp. 116–129). Springer Nature.

Rogol, A. D. (2023). Sex chromosome aneuploidies and fertility: 47, XXY, 47, XYY, 47, XXX and 45, X/47, XXX. *Endocrine Connections, 12*(9).

Roque Bravo, R., Faria, A. C., Brito-da-Costa, A. M., Carmo, H., Mladěnka, P., Dias da Silva, D., Remião, F., & Oemonom Researchers. (2022). Cocaine: An updated overview on chemistry, detection, biokinetics, and pharmacotoxicological aspects including abuse pattern. *Toxins, 14*(4), 278.

Rutland-Brown, W., Langlois, J. A., Thomas, K. E., & Xi, Y. L. (2006). Incidence of traumatic brain injury in the United States, 2003. *The Journal of Head Trauma Rehabilitation, 21*(6), 544–548.

Sandford, R., Beckett, A., & Giulianotti, R. (2022). Sport, disability and (inclusive) education: Critical insights and understandings from the Playdagogy programme. *Sport, Education and Society, 27*(2), 150–166.

Sanjari Moghaddam, H., Mobarak Abadi, M., Dolatshahi, M., Bayani Ershadi, S., Abbasi-Feijani, F., Rezaei, S., Cattarinussi, G., & Aarabi, M. H. (2021). Effects of prenatal methamphetamine exposure on the developing human brain: A systematic review of neuroimaging studies. *ACS Chemical Neuroscience, 12*(15), 2729–2748.

Schalock, R. L., & Verdugo, M. Á. (n.d.) El concepto de calidad de vida en los servicios y apoyos para personas con discapacidad intelectual.

Schaub, A. C., Vogel, M., Baumgartner, S., Lang, U. E., Borgwardt, S., Schmidt, A., & Walter, M. (2022). Striatal resting-state connectivity after long-term diacetylmorphine treatment in opioid-dependent patients. *Brain Communications, 4*(6), fcac275.

Schinz, D., Schmitz-Koep, B., Tahedl, M., Teckenberg, T., Schultz, V., Schulz, J., Zimmer, C., Sorg, C., Gaser, C., & Hedderich, D. M. (2023). Lower cortical thickness and increased brain aging in adults with cocaine use disorder. *Frontiers in Psychiatry, 14*, 1266770.

Schuster, R. M., Crane, N. A., Mermelstein, R., & Gonzalez, R. (2012). The influence of inhibitory control and episodic memory on the risky sexual behavior of young adult cannabis users. *Journal of the International Neuropsychological Society, 18*(5), 827–833.

Scopetti, M., Morena, D., Manetti, F., Santurro, A., Fazio, N. D., D'Errico, S., Padovano, M., Frati, P., & Fineschi, V. (2023). Cannabinoids and brain damage: A systematic review on a frequently overlooked issue. *Current Pharmaceutical Biotechnology, 24*(6), 741–757.

Shumate, C., Allred, R., Dixon, A., Betancourt, D., Yantz, C., Howell, R., Gandhi, H., Kilburn, M., Lupo, P. J., & Agopian, A. J. (2025). Trends in the prevalence of down syndrome (trisomy 21) in Texas by maternal race/ethnicity and maternal age groups, 1999–2020. *American Journal of Medical Genetics, Part A*, e64109.

Srinivasan, K., Canarte, C., Valientes, S. D., Sanchez-Lara, P. A., & Langston, S. J. (2025). Updates in trisomy 18. *NeoReviews, 26*(12), e820–e834.

Stancu, P., Hentsch, L., Seeck, M., Zekry, D., Graf, C., Fleury, V., & Assal, F. (2025). Neurology of aging: Adapting neurology provision for an aging population. *Neurodegenerative Diseases, 25*(1), 14–20.

Steiner, M., & Saenger, P. (2022). Turner syndrome: An update. *Advances in Pediatrics, 69*(1), 177–202.

Støier, J. F., Konomi-Pilkati, A., Apuschkin, M., Herborg, F., & Gether, U. (2023). Amphetamine-induced reverse transport of dopamine does not require cytosolic Ca2+. *Journal of Biological Chemistry, 299*(8).

Stubbins, R. J., McGinnis, E., Johal, B., Chen, L. Y., Wilson, L., Cardona, D. O., & Nevill, T. J. (2021). VEXAS syndrome in a female patient with constitutional 45, X (Turner syndrome). *Haematologica, 107*(4), 1011.

Suchting, R., Beard, C. L., Schmitz, J. M., Soder, H. E., Yoon, J. H., Hasan, K. M., Narayana, P. A., & Lane, S. D. (2021). A meta-analysis of tract-based spatial statistics studies examining white matter integrity in cocaine use disorder. *Addiction Biology, 26*(2), e12902.

Sun, M., He, Q., Sun, N., Han, Q., Wang, Y., Zhao, H., Li, G., Ma, Z., Feng, Z., Li, T., & Yang, L. (2024). Intrinsic capacity, polygenic risk score, APOE genotype, and risk of dementia: A prospective cohort study based on the UK Biobank. *Neurology, 102*(12), e209452.

Tărlungeanu, D. C., & Novarino, G. (2018). Genomics in neurodevelopmental disorders: An avenue to personalized medicine. *Experimental & Molecular Medicine, 50*(8), 1–7.

Thompson, L. D. (2024). Uncommon fibroinflammatory sinonasal tract lesions: granulomatosis with polyangiitis, eosinophilic angiocentric fibrosis, and Rosai–Dorfman disease. *Surgical Pathology Clinics, 17*(4), 549–560.

Thompson, J. R., Schalock, R. L., Agosta, J., Teninty, L., & Fortune, J. (2014). How the supports paradigm is transforming the developmental disabilities service system. *Inclusion, 2*(2), 86–99.

Tondo, L. P., Viola, T. W., Fries, G. R., Kluwe-Schiavon, B., Rothmann, L. M., Cupertino, R., Ferreira, P., Franco, A. R., Lane, S. D., Stertz, L., & Zhao, Z. (2021). White matter deficits in cocaine use disorder: Convergent evidence from in vivo diffusion tensor imaging and ex vivo proteomic analysis. *Translational Psychiatry, 11*(1), 252.

Tosun, D., Bingöl, İ., Akçay, N., Sarıcalı, Ç. Ö., & Mehmet, M. Ş. (2025). Trisomy 18 and Ambiguous Genitalia: A case report. *Journal of Pediatric and Adolescent Gynecology*.

Tran, D. C., Dang, A. L., Van Nguyen, T. B., Tran, V. A., Nguyen, T. H., Le, T. M., Tran, T. L., Ngo, T. A., Hoang, T. N., Le, P. T., & Ngo, T. T. (2023). Typical morphological features on prenatal ultrasound of fetuses with trisomy 13 (Patau's syndrome). *Journal of Clinical Gynecology and Obstetrics, 12*(1), 8–14.

Tripathi, S., Sharma, Y., & Kumar, D. (n.d.) Unraveling APOE4's role in Alzheimer's disease: Pathologies and therapeutic strategies.

Turcotte, S., Kheroua, S., Brun, G., Gagnon, L., Bustamante, N., Labbé, A., Simard, P., Veilleux, M., Lapointe, M., Nguyen, M. H., & Levasseur, M. (2025). Barriers and facilitators to the social participation of individuals aging with a long-term neurological disability: A scoping review. *Disabilities, 5*(2), 49.

Ung, H., McKeon, G., Jokovic, Z., Parker, S., Vickers, M., Malacova, E., Eriksson, L., & Daglish, M. (2025). Long-term neurocognitive side effects of MDMA in recreational ecstasy users following sustained abstinence: A systematic review and meta-analysis. *Journal of Psychopharmacology*, 02698811251389559.

van de Blaak, F. L., & Dumont, G. J. (2022). Serotonin transporter availability, neurocognitive function and their correlation in abstinent 3, 4-methylenedioxymethamphetamine users. *Human Psychopharmacology, Clinical and Experimental, 37*(1), e2811.

Volkow, N. D., Wang, G. J., Fowler, J. S., & Tomasi, D. (2012). Addiction circuitry in the human brain. *Annual Review of Pharmacology and Toxicology, 52*(1), 321–336.

Wengert, E. R., Tronhjem, C. E., Wagnon, J. L., Johannesen, K. M., Petit, H., Krey, I., Saga, A. U., Panchal, P. S., Strohm, S. M., Lange, J., & Kamphausen, S. B. (2019). Biallelic inherited SCN8A variants, a rare cause of SCN8A-related developmental and epileptic encephalopathy. *Epilepsia, 60*(11), 2277–2285.

Wilson, M. P., & Vilke, G. M. (2021). Excited delirium syndrome: Diagnosis and treatment. In *Behavioral emergencies for healthcare providers* (pp. 167–176). Springer Nature.

World Health Organization. (2007). *International classification of functioning, disability, and health: Children & youth version: ICF-CY*. World Health Organization.

World Health Organization. (2024). *Health equity for persons with disabilities: Guide for action*. World Health Organization.

Xi, Z. X., Peng, X. Q., Li, X., Song, R., Zhang, H. Y., Liu, Q. R., Yang, H. J., Bi, G. H., Li, J., & Gardner, E. L. (2011). Brain cannabinoid CB2 receptors modulate cocaine's actions in mice. *Nature Neuroscience, 14*(9), 1160–1166.

Yang, Z., Klugah-Brown, B., Ding, G., Zhou, W., & Biswal, B. B. (2025). Brain structural differences in cocaine use disorder: Insights from multivariate and neurotransmitter analyses. *Progress in Neuro-Psychopharmacology and Biological Psychiatry, 136*, 111159.

Yücel, M., Solowij, N., Respondek, C., Whittle, S., Fornito, A., Pantelis, C., & Lubman, D. I. (2008). Regional brain abnormalities associated with long-term heavy cannabis use. *Archives of General Psychiatry, 65*(6), 694–701.

Yue, J. K., Yuh, E. L., Elguindy, M. M., Sun, X., van Essen, T. A., Deng, H., Belton, P. J., Satris, G. G., Wong, J. C., Valadka, A. B., & Korley, F. K. (2024). Isolated traumatic subarachnoid hemorrhage on head computed tomography scan May not be isolated: A transforming research and clinical knowledge in traumatic brain injury study (TRACK-TBI) study. *Journal of Neurotrauma, 41*(11–12), 1310–1322.

Zhang, Y. X., Zhu, Y. M., Yang, X. X., Gao, F. F., Chen, J., Yu, D. Y., Gao, J. Q., Chen, Z. N., Yang, J. S., Yan, C. X., & Huo, F. Q. (2023). Phosphorylation of neurofilament light chain in the VLO is correlated with morphine-induced behavioral sensitization in rats. *International Journal of Molecular Sciences, 24*(9), 7709.

Zhou, Z. D., Jankovic, J., Ashizawa, T., & Tan, E. K. (2022). Neurodegenerative diseases associated with non-coding CGG tandem repeat expansions. *Nature Reviews Neurology, 18*(3), 145–157.

Chapter 2
Genetic and Chromosomal Syndromes in Disability

Abstract The importance of genetic and chromosomal abnormalities lies in their interference with normal development, leading to various impairments. In this chapter, genetic and chromosomal syndromes resulting in human disabilities are explained. It investigates the way alterations in genes or chromosomes disrupt normal development to cause other physical, cognitive, and functional impairments. The chapter discusses the different forms of genetic mutation and chromosome abnormalities that are commonly associated with disability. It also emphasizes the significance of inheritance patterns and genetic counselling. This chapter will clarify the way people who have genetic and chromosomal syndromes usually experience challenges in the social environment. These obstacles also cause emotional and financial strains on families taking care of them.

Keywords Genetic disorders · Chromosomal abnormalities · Developmental disability · Inheritance patterns · Genetic mutations · Genetic counselling · Cognitive impairment

2.1 Introduction

Chromosomes consist of DNA and proteins organized compactly, with each species exhibiting distinct numbers and structures. Chromosomal disorders and syndromes often arise from numerical and structural chromosomal abnormalities, leading to a variety of symptoms, including those impacting the craniofacial area. They are associated with 50% of spontaneous abortions, 6% of stillbirths, approximately 5% of couples with a history of two or more miscarriages, and roughly 0.5% of babies (Luthardt & Keitges, 2001). Multifactorial inherited diseases represent the primary classification of genetic disorders. These illnesses are not related to Mendelian inheritance patterns; instead, intricate gene-gene interactions and gene-environment interactions substantially affect their development and clinical manifestations. Humans have a total of 23 chromosomes, comprising 22 pairs of autosomes and a

R. Malviya, S. Rajput, *Neurogenetic and Neurodevelopmental Disabilities*, SpringerBriefs in Modern Perspectives on Disability Research, https://doi.org/10.1007/978-981-95-9223-4_2

single pair of sex chromosomes, which are designated as X and Y. Genetic diseases can result from abnormalities in the chromosomes that create an imbalance between them, leading to the acquisition or loss of genetic material (Shaffer & Lupski, 2000). These malformations manifest in live births at a rate of approximately 0.6 and are linked to dysmorphism, structural anomalies, and developmental difficulties. The resulting phenotypes arise from the disruption of dosage-sensitive genes, leading to significant consequences: 25% of miscarriages and stillbirths, along with 50–60% of losses occurring in the first trimester (Zhang et al., 2021).

Numerical aberrations, or aneuploidies, are caused by the loss or gain of entire chromosomes and are often due to an error in segregation during meiosis. They have the highest prevalence of cytogenetic abnormalities and occur with increased tolerance in sex chromosomes and some of the autosomes. Turner syndrome is the best example of a monosomy in relation to the X chromosome (45, X). Trisomy, which is the presence of three copies of a given chromosome, is the case of Down syndrome (trisomy 21), an occurrence of approximately 1 in 800 live births, and other widespread trisomies (exemplified by trisomies 13 and 18). Mosaicism refers to the presence of both normal and defective cell lines, which can lead to certain trisomies, and this phenomenon is most commonly observed in diseases such as those associated with chromosomes 8 and 9. The structural chromosomal abnormalities are caused by the breakage of chromosomes and the reconnection of the chromosome segments, resulting in the loss or gain of chromosome material. The deletions can be classified as either terminal or interstitial, resulting from chromosomal breaks that cause the loss of distal regions or segments situated between pairs of breaks. Duplication, on the other hand, is the acquisition of a segment of a chromosome, whether in its normal or inverted form.

Additional structural chromosomal abnormalities are inversions, translocations, and insertions and do not involve the gain or loss of genetic material. Inversions occur due to two-break events, which invert the intervening chromosomal section. Translocations involve the transfer of part of two chromosomes, whereas insertions occur when part of one chromosome is inserted into a separate region. These rearrangements could be harmful in case they interfere with genes, form new fusion genes, or interfere with the expression of the adjacent genes. Unlike in single-gene mutations, chromosomal disorders interfere with multiple genes that are critical in development. These imbalances can alter the dosage in the affected chromosomal regions that cause diverse clinical outcomes (Table 2.1). The particular outcome results in contiguous gene syndromes, in which many genes in the location that are affected contribute to the clinical phenotype with specific characteristics (Meyyazhagan & Di Renzo, 2023). On the other hand, loss or duplication of a single gene can cause certain chromosomal disorders, which might have pleiotropic outcomes. The deletion issues normally cause haploinsufficiency, where a single working copy of a gene is not sufficient to produce sufficient gene products to enable proper functioning. Syndromes linked to haploinsufficiency encompass Williams syndrome, Miller-Dieker syndrome, Langer-Giedion syndrome, and DiGeorge/velocardiofacial syndrome.

Table 2.1 Genetic and chromosomal syndromes in disability

Syndrome/disorder	Genetic basis	Key neurodevelopmental features
9p trisomy syndrome	Partial duplication of chromosome 9p	Global developmental delay, hypotonia, intellectual disability
Down's syndrome (trisomy 21)	Full or partial extra chromosome 21	Mild–moderate intellectual disability, motor delay, speech delay
Edward's syndrome (trisomy 18)	Extra chromosome 18	Severe cognitive impairment, brain malformations
Mosaic trisomy 22	Mosaicism involving an extra chromosome 22	Developmental delay, motor impairment, variable cognitive disability
Pallister–Killian syndrome (PKS)	Mosaic tetrasomy 12p	Severe hypotonia, developmental delay, intellectual disability
Patau syndrome (trisomy 13)	Extra chromosome 13	Profound intellectual disability, structural brain defects
13q deletion syndrome	Deletion in long arm of chromosome 13	Developmental delay, intellectual disability, speech delay
18q deletion syndrome	Deletion in long arm of chromosome 18	Cognitive impairment, delayed myelination, hypotonia
Angelman syndrome (AS)	Loss of maternal UBE3A gene (15q11–q13)	Severe speech impairment, ataxia, intellectual disability
Cri Du chat syndrome (CdCs)	Deletion of 5p	Severe developmental delay, microcephaly, cognitive impairment
Prader–Willi syndrome (PWS)	Loss of paternal genes (15q11–q13)	Mild–moderate intellectual disability, behavioural issues, hypotonia
Velocardiofacial syndrome (22q11.2 deletion, VCFS)	Deletion at 22q11.2	Learning disability, speech delay, psychiatric vulnerability
Williams syndrome (WS)	Deletion at 7q11.23	Mild–moderate intellectual disability, strong verbal skills, visuospatial deficits
Wolf–Hirschhorn syndrome (WHS)	Deletion of 4p	Severe developmental delay, seizures, intellectual disability
Females with multiple X chromosomes (XXX, XXXX)	Extra X chromosome (s)	Learning disabilities, language delay, motor delays
Turner syndrome (TS)	Monosomy X (45, X)	Mild neurocognitive issues, spatial deficits, motor coordination problems
Klinefelter syndrome (KS)	Extra X chromosome (47, XXY)	Language disorders, learning difficulties, executive dysfunction
XYY syndrome	Extra Y chromosome (47, XYY)	Learning disabilities, language delay, and reduced attention
Mendelian disorders	Single-gene mutations (dominant/recessive)	Variable cognitive impairment, depending on the disorder
Segmental Aneusomy syndromes	Deletion/duplication of small chromosomal segments	Developmental delay, behavioural and cognitive deficits
Aneuploidy (general)	Abnormal number of chromosomes	Wide spectrum of neurodevelopmental disability

In the past, chromosomal abnormalities were identified through microscopic analysis, whereby some of the chromosomal regions (bands) were identified based on their staining intensity. Early studies of chromosomal rearrangements found preliminary evidence of chromosomal rearrangements, but only with the development of sophisticated staining methods were changes of numbers large enough or mutations sufficiently large to be detected. The application of banding methods and high-resolution chromosomal analysis greatly improved the capacity to detect small-scale rearrangements impacting one or more bands (Antonarakis, 2022; Doig et al., 2023). Moreover, molecular cytogenetic techniques like fluorescence in situ hybridization (FISH) and microarrays have significantly advanced the field by enabling the identification of cryptic or sub-microscopic imbalances that remain undetectable through conventional microscopy methods. Such improvements have brought about the identification of other unknown microdeletion and microduplication diseases (Jacquin et al., 2023; Ding et al., 2025). In 1986, Schmickel articulated the concept of contiguous gene syndromes (CGS), defining them as conditions marked by a significant cluster of adjacent genes located on a chromosome (Chen et al., 2024). The term has now been broadened to include the diseases caused by deletions or duplications of chromosomal segments that harbour numerous disease genes, each of which independently affects the phenotype. CGS have been seen in various disorders that are associated with some chromosomes, which usually have very subtle changes in cytogenetics that would not be seen in standard approaches. Advances in molecular cytogenetics, namely FISH and microarray analysis, have played a critical role in the characterization of various CGS. Additionally, cognitive disability, which was once mental retardation (MR), occurs among approximately 3% of the population and has been associated with several social and medical issues. Today, there are not many effective treatments, as they primarily help caregivers to cope with related behavioural issues. The syndrome is highly multifaceted, having both various social and biological factors. Even with the current developments in genetic studies, the knowledge regarding the genetic anomalies that cause MR is still unsatisfactory, making it impossible to enhance care or the understanding of the causes of the condition, which poses a significant obstacle to the field of medicine (Olsen et al., 2023; Zoyirov & Indiaminova, 2021).

2.2 Extent of Genetic Involvement

The severity of the condition is correlated with the aetiology of impaired cognition. Moderate-severe intellectual disability (IQ < 50) is mainly linked to a distinct pathogenic cause (genetic or environmental), yet mild intellectual disability (IQ 50 to 70) may arise from the interplay of various factors. Moderate-to-severe cases are attributed to chromosomal/genetic issues (30–40) and environmental factors (another 10–30). The aetiology is unknown in approximately 40% of the cases (Hirvikoski et al., 2021; Laxova et al., 1977). In cases of mild intellectual disability, the contribution of hereditary and environmental variables is about 30 each, and the remaining 70 do not have a definite aetiology (Rajput et al., 2023). When analysing the pathophysiology of intellectual disability, one should remember that the phenotype involves several areas of intellectual functioning, which are normally categorized into linguistic and performance abilities and memory and attention abilities. Nevertheless, there is still no clear understanding of whether these psychological groupings corroborate genetic factors, particularly in the genetic distinction of verbal and performance skills. Studies of the genetic effects on intellectual functioning are scant and have concentrated on the effects on intellectual functioning in general, such as IQ. The latest research in the field of speech and language development shows that there are certain genetic factors (Pomohaibo et al., 2021; Lewis, 2021). Lacking comprehensive psychometric studies on people with intellectual deficiency caused by certain genetic lesions, it remains unclear whether their cognitive functioning is abnormal or peculiar, though some data suggest the latter.

Molecular cloning and genetic mapping technologies have made possible the exploration of the disorders associated with intellectual disability due to genetic aberrations, especially where structural brain abnormalities cannot be observed. The intricacy has given rise to the distinction of mental retardation (MR) as syndromic and no syndromic. Syndromic mental retardation is identified by the fact that there are concomitant physical defects, including either facial dysmorphism or subtle malformations of the hands and feet, whereas non-syndromic mental retardation is purely an impairment of the mind. Genetic disorders may seem to result in cognitive impairment more frequently in non-syndromic instances than in syndromic ones; nonetheless, this remains a conjecture without a comprehensive explanation of the underlying cause. Identical gene mutations can lead to both conditions, as illustrated by RSK2, which is responsible for Coffin-Lowry syndrome and non-specific intellectual disability (Di Stazio et al., 2021; Wang et al., 2024). Furthermore, alterations in different regions of the ATRX gene can result in both syndromic and non-syndromic forms of intellectual disabilities. Recent progress in comprehending X-linked non-syndromic intellectual disability encompasses the identification of genes that may affect cognition via central nervous system development.

2.2.1 Terminal Rearrangements

Telomeres, situated at the termini of linear chromosomes, are complexes composed of proteins and DNA. They play a crucial role in safeguarding chromosomes from disintegration and fusion, while also ensuring proper pairing, recombination, and segregation of homologous chromosomes during meiosis. The majority of recognizable cytogenetic abnormalities are found at the telomeric ends of chromosomes (Červenák et al., 2021). The telomeric DNA is composed of repeats of G-rich sequences that have been conserved during the evolution of eukaryotes, and human chromosomes end with 2–20 kb of the repeat (TAGGG) (Lukhtanov, 2022). Next to this is the subtelomeric DNA, which is a complex region that spans several hundred kilobases and which has been observed to have polymorphism and dense gene content (Zhou et al., 2022). The subtelomeric regions found on nearly all chromosomes other than acrocentrics 1315 and 2122 have been associated with serious clinical implications, especially idiopathic intellectual impairment, because of their vulnerability to rearrangements that could result in abnormalities (Swierkowska-Janc et al., 2025; Papenhausen et al., 2021). A variety of genetic disorders are characterized by terminal deletions, including the distal 5p deletion associated with cri du chat syndrome and the distal 4p deletion linked to Wolf-Hirschhorn syndrome. Monosomy 1p36 represents the primary terminal deletion syndrome, impacting approximately 1 in every 5000 infants (Papenhausen et al., 2021). The utilization of FISH probes targeting human sub-telomeres has significantly improved the detection of sub-microscopic chromosomal abnormalities among those exhibiting idiopathic intellectual disability without apparent syndromic characteristics (Jacquin et al., 2023). A comprehensive analysis of 11,688 cases where the subtelomere FISH was used revealed pathogenic defects in 2.6% of the subjects (Cui & Li, 2022). Evidence recorded shows that deletions of all telomeric bands are observable (Sun et al., 2024). Chronic abnormalities have been identified and could be correlated with specific phenotypes by comparing the clinical features of affected patients (Phelan et al., 2021). On the other hand, other subtelomeric malformations have been reported in a limited number of cases and require confirmation of similar clinical features to be considered syndromic.

2.2.2 *Recurrent Genomic Structure Mediated Abnormalities*

Various persistent chromosomal anomalies arise through nonallelic homologous recombination (NAHR), which is enabled by neighbouring segmental duplications (Schuy et al., 2022). This approach may result in erroneous crossing-over of nonallelic, homologous regions, yielding two distinct outcomes: tandem or direct duplication and deletion. Genomic disorders are syndromic recurring syndromes with particular criteria: they are typically characterized by rearrangements with segmental duplication breakpoints, they tend to be de novo in individuals who have the disorder, and they share similar symptoms with other patients who have the same rearrangements (Schuy et al., 2022). The genomic structure of established diseases typically features a significant region, varying in size from 50 kb to 10 Mb, bordered by extensive, highly homologous segmental duplications that promote non-allelic homologous recombination (NAHR). NAHR is associated with recurrent chromosome rearrangements in various diseases, including Charcot-Marie-Tooth disease and hereditary neuropathy with pressure palsies vulnerability. This phenomenon has also been connected to several syndromes, such as Angelman, Prader-Willi, and Williams syndromes (Szczawińska-Popłonyk et al., 2023). The magnitude of a recognized anomaly typically remains stable; for instance, the common deletion associated with Williams syndrome is approximately 1.6 Mb, occurring in over 90% of affected individuals (Kozel et al., 2021). Smith-Magenis syndrome is typically characterized by a 5 Mb deletion at 17p11.2, and nearly 90%

of persons with DGS/VCFS have a 3 Mb deletion at 22q11.2, which is sometimes supported by low copy repeats (Zhang et al., 2021). Some non-standard rearrangements do not have the typical segmental duplications and genomic characteristics as were observed in some instances of the Smith-Magenis syndrome and in specific breakpoints in 16p11.2p12.2 microdeletion syndrome (Vervoort & Vermeesch, 2022).

The phenomenon of nonallelic homologous recombination suggests that reciprocal duplications resulting from low-copy repeat deletions should occur with equal frequency; nonetheless, their actual occurrence is less frequent than anticipated (Schuy et al., 2022). The mismatch could be caused by the fact that the people with duplications usually have milder symptoms, thus not requiring professional assessment (Cardoso & Henriques, 2024). Also, small duplications (below 1.5 Mb) can be ignored (Lühmann et al., 2023). Recent population studies based on microarray analysis have found a greater rate of reciprocal duplications than earlier cytogenetic studies (Lindstrand et al., 2022). The MECP2 gene has been reported to experience duplications in boys with developmental delays and in combination with other microdeletion pathologies (Collins & Neul, 2022). The clarity of the situation is uncertain; some of these reciprocal duplications are inherited from the carrier parents, and the phenotypes of the affected individuals may be influenced by unknown genetic modifiers (Kikas et al., 2025). They are different emerging diseases whereby a parent passes them on to his or her children (Theisen & Shaffer, 2010). The microdeletion that occurs at 16p12.1, often in people with idiopathic intellectual disability and congenital anomalies, depicts the intricacies of genetic disease aetiology. It predominates among the affected individuals compared to the normal controls and is normally inherited by the carrier parents, who often exhibit associated diseases like learning disabilities, depression, bipolar disorder, and seizures. The diverse levels of learning impairments observed in adult family members suggest that certain impairments may be pathological, potentially influencing phenotypic expression significantly (Girirajan et al., 2010).

2.3 Chromosomal Abnormalities

The majority of chromosomal defects arise in the early stages of gametogenesis and fertilization, causing the deficiency to show up in all of the organism's cells. Chromosomal abnormalities may arise from changes in the number or structure of chromosomes. The primary causes of numerical aberrations are aneuploidy, which arises from either nondisjunction or anaphase lag, and mosaicism. Embryos with autosomal monosomies as well as trisomies seldom survive due to substantial genetic material loss; however, those with sex chromosomal abnormalities occasionally do, but they demonstrate a variety of developmental issues (Kumar et al., 2015). Structural defects predominantly occur spontaneously due to the destruction or rearrangement of chromosomal material. Various disability

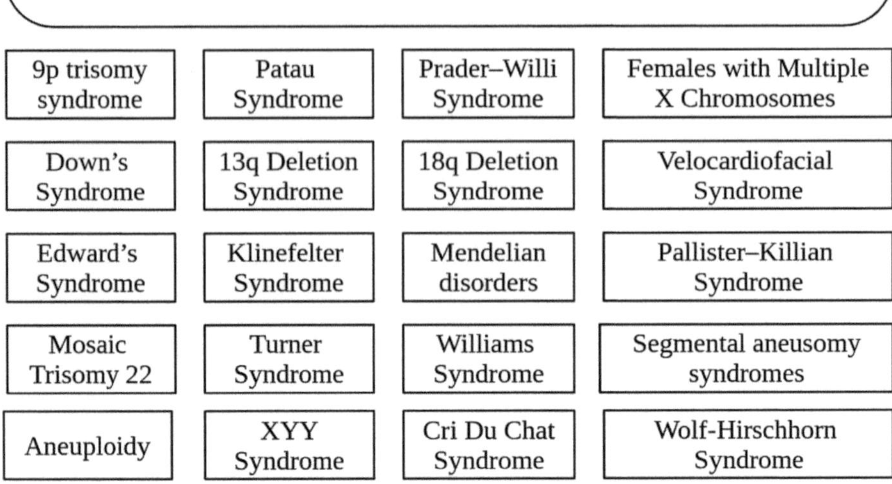

Fig. 2.1 Genetic and chromosomal syndromes in disability

arise due to chromosomal abnormalities are described below (Fig. 2.1 and Table 2.1).

2.3.1 9p Trisomy Syndrome

People identified as having trisomy 9p syndrome, primarily affecting females, have a wide array of phenotypic traits. Initially characterized by Rethore et al. in 1970, more than 125 cases have been recorded in medical literature. The manifestation of the condition differs based on the presence of partial or complete trisomy of the short arm of chromosome 9. Typical characteristics encompass intellectual deficiencies, speech impairments, hypoplasia and dysplasia of terminal phalanges, a singular palmar crease, cyanosis of the extremities, and dysplastic, claw-like nails. Craniofacial anomalies may comprise a prominent, wide forehead, mild micro-brachycephaly, flat occiput, large fontanelle, open metopic suture in childhood, deeply set small eyes, and horizontally or down-slanted palpebral fissures. Supplementary facial characteristics may encompass modest hypertelorism, a prominent nose with a globular tip, downturned nares, a broad mouth with downturned parts, an everted lower lip, micrognathia, a thin, high-arched palate, and microdontia. Common dental disorders include mandibular retrognathism, delayed eruption, and dental crowding. Additionally, some instances of cleft lip and/or cleft palate have been documented (Pisano et al., 2022).

2.3.2 Edward's Syndrome (ES)

Trisomy 18, often known as Edwards syndrome, is the second most prevalent trisomy behind Down syndrome, highlighting a numerical aberration. Esophageal atresia (EA) is far more prevalent in females, occurring in around 1 in 8000 live births, with its incidence increasing with maternal age. The disease primarily arises from an imbalanced translocation of the long arm of chromosome 18, resulting in a lower frequency of mosaicism. Trisomy 18 is associated with low birth weights, deformities of extremities, including webbed or fused toes, and severe intellectual defects. Most of the affected newborns die after 14 days. The known congenital anomalies occur in different systems, such as the fingers, eyes, lungs, diaphragm, heart, and vascular tissues, and renal malformations may also be considered. Additional craniofacial characteristics include microcephaly, low-set ears, and ear malformations. Oral manifestations may include microstomia, retrognathia, micrognathia, and a thin palate (Carey, 2021).

2.3.3 13q Deletion Syndrome

The 13q deletion syndrome represents a rare chromosomal anomaly, with approximately 120 documented instances in the medical literature. This condition arises from a deletion that results in the development of a ring chromosome. This variant is categorized into three distinct groups according to the location of the deletion breakpoint: group 1, which encompasses deletions proximal to q32; group 2, which includes q32; and group 3, which pertains to the regions 13q33 and 13q34. The estimated prevalence rate stands at 1 in every 58,000 live births. Typical features include cognitive impairment, a short neck accompanied by redundant skin folds, congenital heart defects, hypospadias, imperforate anus, bifid scrotum, cryptorchidism, abnormalities of the pelvic girdle, and malformations of the feet and toes, such as absent thumbs and shortened fourth and fifth fingers. Craniofacial abnormalities include a prominent nasal bridge, hypertelorism, microcephaly, epicanthus, ptosis, microphthalmia, retinoblastoma, abnormally positioned low-set ears, colobomata, and facial asymmetry. Oral features encompass micrognathia and a protruding maxilla (Efimova et al., 2025; Kaya et al., 2022).

2.3.4 Females with Multiple X Chromosomes

The disease is unique to females and is associated with additional X chromosomal copies or mosaicism, which occurs among approximately 1 in 1000 people. Typical features are behavioural and emotional problems, the probability of learning impairments, and speech, motor, and language delays (Talebizadeh et al., 2006).

Craniofacial features include a hypertelorism, large flat nose, and epicanthus, but oral features similar to those of Down syndrome may be present, including big jaws.

2.3.5 *Angel Man Syndrome (AS)*

In 1950, a researcher first described Angelman syndrome (AS), which was initially called Happy Puppet Syndrome. Microdeletions represent the predominant etiological factors in Angelman Syndrome (AS), accounting for 70% of cases. Other contributing factors include uniparental disomy, which is observed in 5% of instances, and aberrant methylation in the 15q11-13 region resulting from imprinting mutations, found in 4% of cases. Its occurrence rate is about 1 per 10,000 live births. Such clinical manifestations are severe intellectual disability, the lack of verbal communication, the abnormalities of the electroencephalogram, ataxia, the rigid gait, developmental delays, aggressive behaviour, inappropriate laughter, and seizures. The facial characteristics encompass microcephaly and deep-set eyes, while the oral attributes may consist of macrostomia, maxillary hypoplasia, mandibular prognathism, a projecting tongue, and widely spaced teeth (Duis et al., 2022).

2.3.6 *Prader-Willi Syndrome*

Prader-Willi syndrome (PWS) is a genetic condition linked to paternal deletions on chromosome 15, primarily resulting from microdeletions in 70% of cases and uniparental disomy in 30 per cent. Prader-Willi syndrome was first described in 1956. It is equally prevalent between the sexes, and the rate of incidence is 1:12,000 to 15,000. Typically, the prevalent characteristics are increased susceptibility to emotional, behavioural, and cognitive problems. Possible preliminary signs may include loss of foetal activity, lack of muscular tone, incomplete development of the genitalia, retarded skeletal maturation, and delayed development. Oral manifestations encompass dental caries, enamel hypoplasia, microdontia, malocclusion, a significant accumulation of calculus, reduced salivation, gingivitis, and a distinctive fish-like mouth characterized by a triangular upper lip and arched palate (Muscogiuri et al., 2021).

2.3.7 *WHS (Wolf-Hirschhorn Syndrome)*

WHS was first characterized in 1961 and is linked to deletions in the 4p chromosomal region, primarily resulting from de novo deletions. Unbalanced translocations result in a minority, whose severity is related to the extent of deletion. WHS is mostly a disease affecting the female gender, with its incidence rate of between 1 in

20,000 and 1 in 50,000 births. Common characteristics encompass developmental retardation, intellectual retardation, anomalies of midline fusion, behavioral/intellectual discrepancies, seizures, and sleep disturbances. Other serious symptoms encompass the lack of muscle development and deformity of several organs. Craniofacial appearances are similar to those of a Greek warrior helmet, with a large forehead and a large gap between the eyes, and oral defects are a cleft lip or palate (Gavril et al., 2021).

2.3.8 *Cri Du Chat Syndrome (CdCs)*

CdCS, or 5p syndrome, or cat-cry syndrome, is one of the earliest known genetic diseases. The condition is characterized by a characteristic high-pitched feline-like vocalization with a covalent relationship of chromosome 5p15.3 metabolomic pathways; approximately 90% of the cases are a result of the missing short arm of chromosome 5, and 10% by unbalanced translocations in families. The prevalence rate of CdCS stands at approximately 1 in 50,000 live births. The predominant clinical characteristics encompass developmental delays, language delays, and cognitive delays, and the symptoms vary depending on the extent and location of the deletion in chromosome 5p. Additional related symptoms may encompass behavioural challenges, diminished cognitive abilities, auditory impairments, and scoliosis; however, a minority of those affected might also exhibit notable organ anomalies. Craniofacial characteristics include microcephaly, widely spaced eyes, a broad nasal bridge, and oral manifestations such as micrognathia, malocclusion (notably overjet), and cleft lip and/or cleft palate (CLCP) (Papadopoulou et al., 2024).

2.3.9 *18q Deletion Syndrome*

The 18q deletion syndrome is a common genetic disorder, affecting around 1 in 10,000 live births, initially identified by de Grouchy et al. in 1964. The phenotypic expressions are contingent upon the degree of the deletion on the long arm of chromosome 18, particularly within the 18q22.3 region. This syndrome is defined by intellectual incapacity, diminished stature, hypotonia, auditory deficits, foot deformities, and other endocrine disorders, sometimes associated with autoimmune diseases. Craniofacial characteristics include microcephaly, midface hypoplasia, a shortened palpebral fissure, deep-set eyes, a carp-shaped mouth, and external ear malformations, such as preauricular pits and a constricted or atretic external auditory canal. Additionally, oral manifestations may encompass palatal anomalies (Liu et al., 2021).

2.3.10 Velocardiofacial Syndrome (VCFS)

VCFS, which is an abbreviation of "vertebral palatal", "cardiac", and "facial", is a chromosomal microdeletion disease, which is associated with genetic changes on the long arm of chromosome 22, specifically affecting the SHANK3 gene. It equally affects both sexes, with an incidence rate of about 1 in 2000 to 1 in 4000 among newborn babies. VCFS is linked to a range of psychiatric disorders, including schizophrenia, attention deficit disorder, and bipolar disorder. It is also associated with learning disabilities and developmental delays, as well as cardiovascular and ocular issues. Additionally, individuals may experience frequent occurrences of otitis media, immunodeficiencies, hypocalcemia, scoliosis, and skeletal deformities of the upper back. Major craniofacial features are the result of long face structures, almond-shaped eyes, eyelashes, cheeks, and odd ears. About seven out of ten VCFS patients exhibit velopharyngeal insufficiency caused by conditions such as cleft palate (CP), which is characterized by platybasia, adenoid hypotrophy, enlarged tonsils, hypotonia, and poor muscle activity of the pharynx. Soft palate clefts are often accompanied by oral symptoms (McDonald-McGinn et al., 2021).

2.3.11 Turner Syndrome (TS)

Turner syndrome (TS) is an X-linked condition marked by the complete or partial absence of one X chromosome in females, affecting roughly one in every 8000 female births. Half of the people with Turner syndrome have a 45, X0 karyotype, and others can be found to be mosaic or with different X chromosomal abnormalities. Turner Syndrome is predominantly found in the development of females and has no foundation with maternal age, although in 75% of cases, the syndrome is characterized by the absence of the X chromosome of the male sex. The individuals affected are usually infertile, webbed, skeletal, and scoliosis and prone to various health problems, including heart diseases and osteoporosis. There is occasionally a slight dysmorphic craniofacial appearance, such as a small chin and curled upper lip, and oral appearances of micrognathia and malocclusion (Gravholt et al., 2023).

2.3.12 Patau Syndrome (PS)

Patau syndrome, first identified by Dr. Klaus Patau in 1960, represents the rarest form of primary autosomal trisomy, manifesting in approximately 1 in 20,000 live births, with its incidence increasing alongside maternal age. The survival rate after infancy is rare, and about 70 per cent of the cases do not survive after 6 months. The disorder is mainly attributed to meiotic errors in the maternal and less often by mosaicism. The most prevalent congenital malformations that correlate with PS are severe respiratory and heart problems, myelomeningocele, and optic and olfactory nerve hypoplasia, as well as multiple ocular anomalies like holoprosencephaly. Other complications might involve hearing issues, genital problems, scoliosis, and craniofacial anomalies such as microcephaly, flattened facial appearance and palate, or clefting issues like micrognathia and cleft lip or palate (Piao et al., 2022).

2.3.13 Pallister–Killian Syndrome (PKS)

Pallister-Killian Syndrome (PKS) was initially documented in 1977 and then separately described in 1981, attributed to a tetrasomy of the chromosomal region 12p. Approximately 100 cases have been recorded in medical literature. Typical clinical manifestations in neonates and children with PKS encompass hypotonia, pronounced cognitive and motor impairment, seizure disorders, and frontotemporal alopecia, although less often observed individuals may exhibit hyper- or depigmentation. Associated abnormalities may include bone deformities such as rhizomelic brachymelia, alongside ocular and cardiovascular complications, genital malformations, and congenital diaphragmatic defects. Craniofacial characteristics include features such as raised frontal hairlines, hypertelorism, a broad and flat nasal bridge, low-set dysplastic ears, frontal bossing, exophthalmos, shallow upper orbital ridges, upward-slanting palpebral fissures, inner epicanthic folds, diminutive upturned nares, prominent cheeks, an enlarged and simple philtrum, a pronounced upper lip, and short necks frequently associated with webbing and excessive nuchal skin. Oral manifestations can present as macrostomia characterized by downturned corners, an enlarged mandible, high-arched palates, and instances of cleft lip and palate (CLCP), in addition to labial pits observed on the lower lip (Arghir et al., 2021).

2.3.14 Klinefelter Syndrome (KS)

Klinefelter syndrome, also known as XY syndrome, arises from the presence of an additional X chromosome in males, with mosaic forms accounting for approximately 10% of the instances. The rate of occurrence is about 1 per 1000 genders in males. People are not always in the knowledge of their sickness until the onset of puberty. KS is associated with several health challenges, such as malformation in the genitals, high risk towards breast cancer, infertility, osteoporosis, respiratory conditions, autoimmune diseases, and cognitive and motor disabilities. The oral symptoms might include maxillary and mandibular prognathism, bigger permanent tooth crowns, and taurodontism (Blackburn et al., 2025).

2.3.15 Down's Syndrome (DS)

Down syndrome, the most prevalent chromosomal condition, is characterized by the presence of an additional chromosome 21 resulting from trisomy, affecting around 1 in 700 to 1 in 800 live births. The majority of the cases (94%) are a result of meiotic nondisjunction, with unbalanced Robertsonian translocation (4%) and mosaic (2%) cases happening. People affected by Down syndrome show a spectrum of general traits since birth, such as learning disabilities, different degrees of intellectual disability and neurodevelopmental anomalies, and cardiovascular conditions. The most significant skeletal anomalies comprise elongated limbs, pelvic dysplasia, clinodactyly, a space between the first and second toes, and instability of the atlanto-occipital joint. The typical symptoms are visual defects, transverse pitting creases, hypogonadism, abdominal distension, and delayed puberty. The craniofacial appearances are peculiar, with brachycephaly, flat occiput, full neck, flat facial profile with hypertelorism, strong epicanthic folds, maxillary hypoplasia, and outwards-tipped palate fissures. In the case of the affected individuals, oral manifestations may be severe periodontal disease, fewer tooth decays, malocclusion, late permanent eruption of teeth, microdontia, congenital tooth absence and malformation, thickened or fissured lips, macroglossia, angular cheilitis, and such problems as bruxism, tongue thrusting, and mouth breathing (Schmidt et al., 2022; Lovras et al., 2025).

2.3.16 Williams Syndrome (WS)

Williams syndrome (WS) represents a unique neurodevelopmental disorder characterized by a microdeletion of neighbouring genes on chromosome 7, notably involving the elastin gene (ELN) alongside cardiovascular anomalies. The incidence of the syndrome ranges between 1 in 20,000 and 1 in 50,000. Williams syndrome is portrayed in the early stages of life as failure to thrive, developmental delay, heart defects, and other physical defects like hernias, esotropia, chronic otitis media, skeletal defects, kidney complications, and hypercalcemia. During adulthood, the effect is urinary tract infections, peptic ulcers, cholelithiasis, and gastrointestinal problems. Craniofacial defects comprise dolichocephaly, cranial asymmetry, and/or nasal and oral structures, e.g. hypodontia, microdontia, and enamel hypoplasia, frequently causing dental complications, e.g. malocclusion and increased prevalence of dental caries (Morris & Mervis, 2021).

2.3.17 *XYY Syndrome*

The XYY syndrome is a disorder that occurs in males and is a result of having an extra X chromosome, denoted 47, XYY. Mosaic can also be due to the dependence of phenotypic effects on the XY/XYY ratio of cells. Most individuals with XYY syndrome experience normal lives and frequently remain unaware of their condition. However, there are minor issues that individuals can experience, e.g. speech difficulties, learning disorders, and behaviour disorders, including anger management and aggression issues. Some examples include oral signs, including abnormally large deciduous and permanent dentures, as well as shovel-shaped lateral incisors (Patil et al., 2014).

2.3.18 *Mosaic Trisomy 22*

Trisomy 22 is a rare chromosomal defect that is caused by the extra presence of chromosome 22. It appears in various forms, such as a mosaic form, which is viable, but total non-mosaic trisomy 22 is accompanied by spontaneous abortions. The common features include growth retardation and extreme intellectual disability, as well as anomalies such as a webbed neck, abnormalities in limbs, and hypomelanosis of Ito. Associated issues can be congenital cardiac defects, hearing loss, and genital disorders, including cryptorchidism. Craniofacial features encompass microcephaly, macrocephaly, a flattened nasal bridge, a large forehead, preauricular pits, bilateral epicanthic folds, hypertelorism, and low-set ears. Oral manifestations may include cleft palate and micrognathia (Trevisan et al., 2024).

2.4 Intellectual Disability

The early onset of intellectual disability (ID), previously referred to as mental retardation (MR), is a significant impairment in the ability to process complex information and act independently; it develops before adulthood and influences the developmental trajectory (Pascual et al., 2023). Intellectual Disability (ID) diagnosis and measurement of the severity are usually done using standardized IQ tests, such as Wechsler Adult Intelligence Scales (WAIS), and the Wechsler Intelligence Scales for Children (WISC), although they are not specifically tailored to that end. Intellectual disability (IQ < 70) is prevalent in Western countries, with an estimated 1.5–2%, and a severe case (IQ less than 50) of 0.3 to 0.5%. Studies indicate that the prevalence is significant in developing nations, attributed not to genetic factors but rather to non-genetic influences such as hunger, cultural deprivation, and inadequate healthcare resources (Yong et al., 2021; Olusanya et al., 2022). Moreover, paternal consanguinity has also been listed as an important etiological risk factor, with inbreeding being linked to reduced cognitive ability (Townsend et al., 2025). Although the intellectual disability presents a significant socio-economic burden, even more than dementia, there is still a lack of awareness of it among the population, as most professionals and organizations often consider intellectual disability a social or educational problem, but not a healthcare issue (Huisman et al., 2024; Samuel et al., 2023).

Research shows that moderate intellectual disabilities (ID) are likely to be located on the low end of normal IQ distributions that are caused by genetic and environmental factors. On the other hand, deep intellectual disability is often associated with major incidents, including perinatal hypoxia, prenatal infections, or such genetic diseases as chromosomal abnormalities (Zedan et al., 2023). The genetic mechanisms of severe intellectual disability have been demystified in recent discoveries, and cytogenetically detectable chromosomal abnormalities are present in about 15% of cases. In addition, minor deletions and duplications, which cannot be detected using conventional karyotyping, are currently identified as noteworthy causes of intellectual deficiency. Malfunctions in X-linked genes account for approximately 10% of cases of intellectual disability in males, indicating that additional factors play a role in the higher prevalence of cognitive impairment observed in males compared to females. Approximately 60% of intellectual disability remains of unknown origin, suggesting the potential presence of an unidentified autosomal gene deficiency. This phenomenon may involve both dominant and recessive mutations occurring sporadically, particularly within the smaller family units characteristic of industrialized nations (Xi et al., 2024; Talon et al., 2021).

2.4.1 Chromosomal Aberrations and ID

In light of the recent progress in prenatal diagnostics for older mothers, Down syndrome (trisomy 21) continues to be the predominant cause of intellectual disabilities (ID) (Rivera et al., 2021). Chromosomal aberrations that can be identified cytogenetically can be observed in about one in seven people with substantial cognitive impairment, but other chromosomal aberrations are less prevalent. The advancement of high-resolution array comparative genomic hybridization (CGH) has led to the identification of small deletions and duplications as significant contributors to intellectual impairment (ID). The application of a BAC array for the detection of copy number variations (CNVs) revealed pathogenic rearrangements in 5.5% of a cohort comprising 290 patients diagnosed with intellectual disability (ID) (Vollger et al., 2022). In the biggest cohort evaluated for CNVs, around 15% of patients demonstrated causal de novo deletions or duplications, primarily non-recurrent and unlinked to low-copy repeats (Brand et al., 2021). This suggests that processes including non-homologous end joining, replication fork stalling, and template swapping contribute to the non-recurrent genomic aberrations seen in conditions such as Pelizaeus-Merzbacher syndrome (Cao et al., 2022). Over the past 2 years, several copy number variations (CNVs) linked to intellectual disability (ID) and developmental delay have been identified, as detailed in reviews by Stankiewicz and Beaudet, among others (Hamad et al., 2023). These findings have substantially expanded the collection of identified microdeletion and duplication diseases. Particular genetic anomalies, especially the microdeletion of the 1p36.1 region, are notably common, impacting around 1% of individuals diagnosed with "idiopathic" intellectual disability (Jacquin et al., 2023).

Array Comparative Genomic Hybridization (CGH) has been useful in identifying functional defects linked to intellectual disability (ID) in many conditions, including Peters plus, CHARGE, and Pitt-Hopkins syndrome, attributable to specific gene abnormalities (Wang et al., 2022). A singular deletion of the MAPT gene is associated with developmental anomalies in people with 17q21.31 microdeletions (Saia et al., 2023). Moreover, mutations that modify TBX1 function are linked to the consequences of prevalent 22q11.2 deletion or duplication disorders (Purow et al., 2025). Prior investigations into mutations have revealed that the haploinsufficiency of the RAI1 gene accounts for the features associated with Smith-Magenis syndrome, while similar findings have been observed concerning the UBE3A gene in Angelman syndrome (Bossuyt et al., 2021). Consequently, Copy Number Variants (CNVs) offer essential insights for detecting gene abnormalities linked to intellectual disability and enhance the comprehension of its complex pathogenesis. Approximately 3% of persons with de novo balanced chromosomal rearrangements have moderate to severe intellectual impairment (ID) (Lowther et al., 2022). Systematic breakpoint mapping has identified multiple candidate genes associated with intellectual impairment (ID). Recent advancements in methodologies, such as chromosome sorting, high-resolution array comparative genomic hybridization (CGH), and next-generation sequencing, have significantly refined the molecular characterization of breakpoint regions and facilitated the identification of truncated genes (Fig. 2.2) (Jansen et al., 2023). Nonetheless, earlier research must account for submicroscopic deletions or duplications present across the genome, which have been noted in particular cases, especially among individuals with notable congenital anomalies (Jansen et al., 2023).

X-linked intellectual disability (XLID) is characterized by its specific inheritance patterns and accounts for 8–12% of intellectual impairment cases in males, even though the X chromosome contains only 4% of the protein-coding genes within the human genome. More than 80 genes connected to X-linked intellectual disability have been found, predominantly in the past decade via extensive family cohort studies. Mutations in 30 of these genes are connected to non-syndromic X-linked intellectual impairment, representing more than 50% of cases. However, additional genes associated with syndromic XLID further complicate the clinical difference between the two types. The introduction of supplementary patient data may facilitate the identification of certain clinical patterns, hence reducing the incidence of non-syndromic patients. Recent investigations involving 900 documented X chromosomal genes across 250 XLID families have revealed the presence of two or more protein-truncating mutations in eight novel genes. Additionally, other sequence variants have been identified; however, their clinical importance remains unclear (Bufton et al., 2022). Sequence variations associated with XLID have been

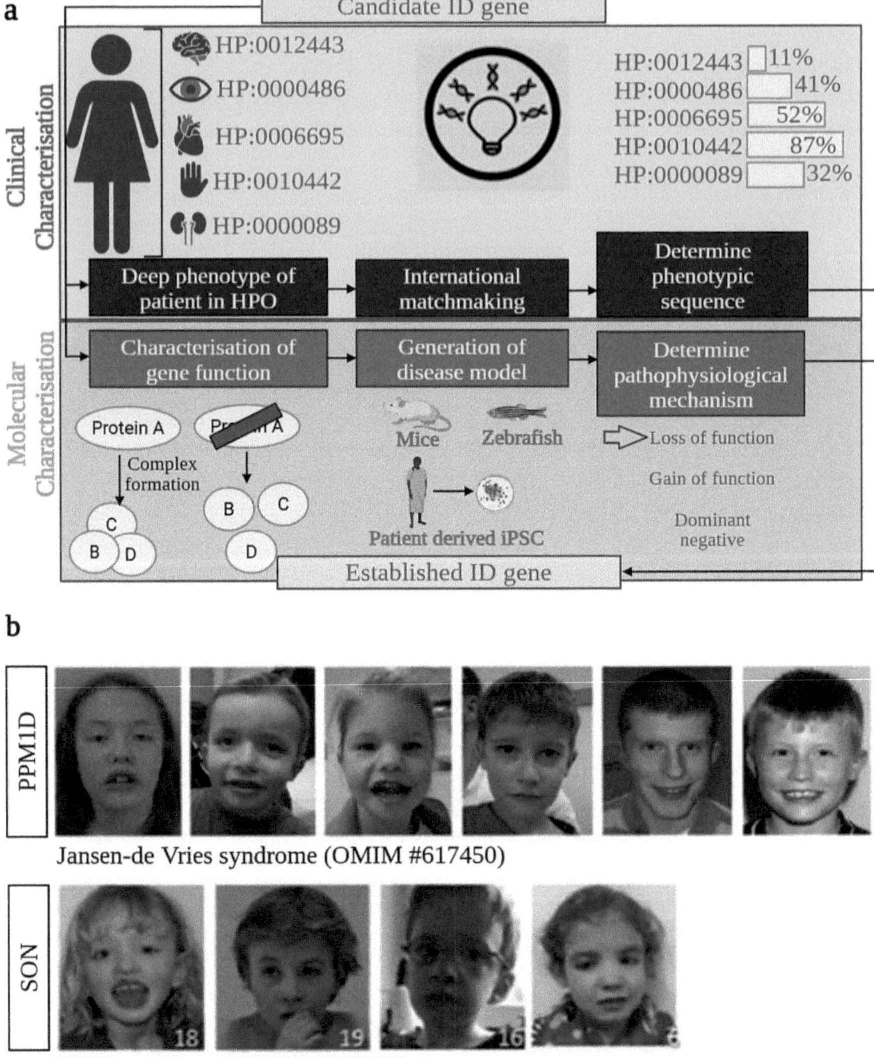

Fig. 2.2 A schematic overview illustrates a procedure for reclassifying a candidate ID gene into a validated ID gene, showcasing two successful examples. (**a**) Exome sequencing is utilized to uncover novel disease-gene correlations, commencing with the identification of a (presumably) harmful variant in an unrelated gene. This initiates two simultaneous investigations: one to examine the clinical characteristics of people with the candidate ID gene, and another to assess the gene's molecular dysfunction. (**b**) Additionally, face scans indicate two neurodevelopmental disorder syndromes identified through exome sequencing: PPM1D associated with Jansen-de Vries syndrome and SON connected to ZTTK syndrome. (Taken from the manuscript Jansen et al. (2023) published under creative coma open access licence)

detected in only a small fraction of families, attributed to many factors, including the potential exclusion of non-coding mutations during exon sequencing or the presence of undetected minute submicroscopic duplications. A prevalent duplication of the MECP2 gene has been identified in roughly 1% of males with moderate to severe X-linked intellectual impairment (XLID) (Pascual-Alonso et al., 2021). Moreover, pathogenic copy number variations (CNVs) greater than 100 kb have been detected in 5% of over 400 families analysed by the European MRX Consortium, suggesting that these changes do not fully explain the "missing" mutations observed in the IGOLD study (Tolmacheva et al., 2022). The composition of smaller families in this study may omit cases with X-linked intellectual disability, aligning with observations in other cohorts.

2.4.2 Autosomal Forms of ID

The acute dominant forms of intellectual impairment (ID) are typically non-hereditary owing to diminished reproductive rates in affected individuals. The prevalence of dominant intellectual disabilities (ID) is little documented; nevertheless, the incidence of de novo copy number variations (CNVs) in individuals with idiopathic ID suggests it is very common. Conversely, most genetic abnormalities leading to intellectual impairment (ID) are likely inherited as recessive features, indicating that autosomal recessive forms of ID (ARID) may be relatively common, potentially stemming from mutations in many genes. The homozygosity mapping technique in extensive consanguineous families is essential for identifying molecular abnormalities linked to ARID, hence aiding in the mutation analysis of candidate genes (Vitale et al., 2022). This methodology has recently led to the identification of many genetic defects linked to autosomal recessive microcephaly and syndromic ARID variations (Manning et al., 2025). Non-syndromic ARID is regarded as more prevalent than syndromic cases; nonetheless, before 2007, only a restricted number of loci and a maximum of three non-syndromic ARID genes were discovered, primarily due to a shortage of extensive families and parental consanguinity in industrialized nations (Higgins et al., 2004). Najmabadi et al. conducted the first extensive mapping of genes linked to non-syndromic Autism-Related Intellectual Disabilities (ARID) using homozygosity mapping in 78 consanguineous Iranian families (Najmabadi et al., 2007). Their findings indicated that ARID is very diverse, making it unlikely that common genetic defects account for a considerable proportion of cases. The investigation reported eight novel loci, including a deletion of the GRIK2 gene, and further research uncovered nine additional valid loci, increasing the total to 22 (Stolz et al., 2021). The bulk of these linkage intervals are unconnected, further corroborating the variability of non-syndromic ARID (Kuss et al., 2011). Recent findings have identified two protein-truncating mutations and a complex deletion mutation in the TUSC3 gene within French and Iranian families. These discoveries affirm that TUSC3 is the sole gene associated with autosomal recessive ARID, with multiple mutations documented in the literature (Molinari et al., 2008;

Garshasbi et al., 2008). Furthermore, a disabling mutation in the FTO gene, which has been previously linked to obesity, has been identified as an additional factor in ARID (Frayling et al., 2007; Malviya & Rajput, 2025). Currently, six genes are linked with non-syndromic ARID, and the ongoing studies have shown that more genes can be discovered in the future. Furthermore, advancements in SNP typing through arrays have enabled the high-throughput mapping of homozygosity in extensive consanguineous families affected by recessive diseases. Innovative methodologies are continuously being developed to markedly expedite the processes of mutation screening and genome discovery.

2.4.3 Syndromic Intellectual Disability

2.4.3.1 Mendelian Disorders

The most recognized Mendelian intellectual disabilities are X-linked, primarily arising from X-linked recessive disorders that ensure the continued presence of affected individuals through generations, thus categorizing them as hereditary conditions and facilitating genetic mapping. X-linked intellectual disability (XLMR) is anticipated to manifest in 1.8 individuals per 1000 males, alongside a carrier frequency of 2.4 per 1000 females. Currently, 210 instances have been recognized as X-linked, with 126 mapped and 32 cloned (Chiurazzi et al., 2001). Fragile X syndrome is the most common type of X-linked mental retardation (XLMR), impacting around 1 in 5000 boys and 1 in 8000 girls. It is characterized by a delicate locus at Xq27.3 linked to a trinucleotide repeat (CGG) expansion in the FMR1 gene. The exact function of FMR1 and the mechanism by which its absence leads to intellectual impairment is still unclear, despite its importance for positional cloning (Jin & Warren, 2000). In a healthy brain, FMR protein is present in nearly all neurons and is associated with RNA binding, as well as possibly with mRNA export and translation processes (Yang et al., 2025). A theory suggests that FMR1 influences dendritic spine formation and pruning during cerebral development, which may be essential for optimal brain function (Song & Broadie, 2023).

The changes in the mechanisms of gene expression have been recognized as important genetic factors causing intellectual impairment. Two different syndromic conditions have been identified with genes that play the role of transcriptional regulators that impact chromatin (Good et al., 2021). Rett syndrome is a disorder that is mainly encountered in women; it is associated with alterations in the methyl-CpG-binding protein 2 (MeCP2), an interaction with CpG dinucleotides, which is used to induce transcriptional repression by interacting with histone deacetylase and the corepressor SIN3A. The abnormalities of the ATRX gene correlate with the alpha-thalassemia X-linked mental retardation syndrome (ATRX), which leads to developmental problems such as intellectual disability, facial dysmorphism, urogenital anomalies, and alpha-thalassemia. The ATRX gene has motifs that indicate binding to chromatin, and mutations in this gene affect the genomic patterns of methylation, which is in line with its role in chromatin remodelling (Scott et al., 2021). The diverse effects of MECP2 and ATRX mutations demonstrate that they might be involved in the regulation of a specific set of gene expressions.

It has been discovered in the exploration of the syndromic mental retardation (MR) disorder called CLS that the MAPK-activated signalling pathway, which is attributed to cognitive deficits, is involved. CLS is marked by severe psychomotor dysfunction, numerous bodily anomalies, and progressive skeletal malformations. The genetic locus linked to this disease has been located in the region Xp22.2, where mutations have been identified in the candidate gene RSK2 (RPS6KA3) (Rogers, 2021). RSK2 is pivotal in promoting the phosphorylation of the cyclic adenosine monophosphate response element-binding protein (CREB) when stimulated by growth factors. This process is essential for a response mechanism that allows the Ras-MAPK and Ras pathways to activate gene expression within the nucleus. Non-syndromic intellectual disability is associated with mutations in RSK2; patients with an X-linked intellectual disability family who lacked physical manifestations of Coffin-Lowry syndrome but had mild intellectual disability showed a mutation in exon 14 of RSK2, and this mutation resulted in a conservative change in amino acids. The exact pathogenesis of this condition is also not clear.

2.4.3.2 Segmental Aneusomy Syndromes

Many genetic pathogeneses of intellectual disability are explained by small deletions or duplications of chromosomes, which are categorized as segmental aneuploidy syndromes (Shaffer, 2021). These are smaller (usually less than five megabases) areas that enable researchers to narrow the dose-sensitive gene. Families with point mutations associated with intellectual disabilities must be discovered to establish the sensitivity of the dose of the gene. Rubinstein-Taybi syndrome is known for unique craniofacial features, broad thumbs, and cognitive impairment, which is a consequence of monosomy of a particular gene located in 16p13.3 (Matasariu et al., 2025). CREB-binding protein (CBP) (Van Gils et al., 2021) is encoded by the gene related to this disease. Williams-Beuren syndrome represents a neurodevelopmental disorder distinguished by congenital heart defects, infantile hypercalcemia, characteristic facial dysmorphia, and intellectual impairment resulting from haploinsufficiency of genes located in the 7q11 locus (Ferrari & Stagi, 2021). The elastin gene is recognized for its role in supravalvular aortic stenosis; however, it is posited that a minimum of 15 additional genes may also play a part in the atypical cognitive profile linked to the syndrome. The possible genes include various transcriptional regulators, among which are the Williams syndrome transcriptional factor, LIMK1, and the WBSCR14/WS-bHLH gene, which are characterized by certain structural elements, such as PHD and bromodomains, which are required in their regulatory functions (Zhou et al., 2022).

The anomalies of the chromosomal region 15q11-q13 link Angelman syndrome (AS), and Prader-Willi syndrome (PWS), two distinct diseases. AS is characterized by significant cognitive impairments, limited verbal communication in the majority of patients, ataxia, abnormal EEGs, seizures, microcephaly, hyperactivity, facial dysmorphism, and inappropriate laughing outbursts. On the other hand, Prader-Willi Syndrome is usually characterized by mild intellectual retardation, characteristic facial appearance, and hyperphagia, which result in excessive obesity. Despite the clinical distinctions, both Prader-Willi syndrome (PWS) and Angelman syndrome (AS) arise from analogous parent-of-origin-specific gene expression deficiencies on chromosome 15. Prader-Willi Syndrome (PWS) occurs when one parent is a mother who inherits both copies of chromosome 15, and Angelman Syndrome (AS) occurs when one parent is a father who inherits both copies. Notably, almost 25% of Prader-Willi Syndrome arises due to the inheritance of two mother chromosome 15s rather than the chromosomal deletions, which are against the normal arrangement of one chromosome that belongs to the mother and the other belonging to the father. On the same note, two paternal chromosomes 15 may be inherited, leading to AS. It is a region of the chromosome that has a parent-of-origin imprint that is manifested by changes in DNA methylation (Zhang et al., 2021). Likewise, some non-characteristic cases of AS are linked to the mutations in the UBE3A gene that plays a vital role in the system of protein degradation via ubiquitin (Bossuyt et al., 2021).

PWS is likely not caused by a single gene mutation, as seven brain-specific genes in the PWS region have been found (Kolbasin et al., 2024). Evidence suggests that the deficiency in PWS may result from aberrant RNA editing caused by the dysregulation of guide RNAs. Within the nucleolus, there is a collection of short nucleolar RNAs (snoRNAs), which mainly take part in posttranscriptional rRNA nucleotide modification. Recent discoveries have identified three brain-specific snoRNAs located within the crucial 15q11 region linked to Prader-Willi Syndrome (PWS) and Angelman Syndrome (AS), exhibiting genomic imprinting in both mice and humans (Gawade & Raczynska, 2024). Their operations are not well understood due to the lack of their proper antisense elements; however, one is associated with an mRNA that encodes the serotonin receptor-2C, which means that it may be related to the abnormalities of serotonin neurotransmission in PWS (Rajput et al., 2024). Similarly, the cognitive impairments associated with the deletion in the 22q11 region, linked to velocardiofacial syndrome (VCFS), DiGeorge syndrome (DGS), and conotruncal anomaly face (CTAF), present a significant challenge. DGS as well as VCFS both have cognitive impairments, and some individuals with VCFS have psychosis (Cortés-Martín et al., 2022). Although the region has at least 14 genes, there has been no conclusive selection of the likely causal factors of cognitive impairments; this could indicate that the syndromes are caused by interactive monosomy of a number of genes.

2.4.3.3 Aneuploidy

The segmental aneuploidies are difficult to analyse, and the identification of specific genes that are associated with aberrations in aneuploidies is also a challenge. The comparative analysis of patients with incomplete aneuploidy has demonstrated the areas of serious domains related to Turner syndrome (XO) and Down syndrome (trisomy 21) (Antonarakis, 1998). Some of the candidate genes associated with the somatic characteristics of Turner syndrome are SHOX/PHOG, which is associated with short stature, and RPS4Y, which is associated with functions of ribosomal proteins (Fisher et al., 1990). Further characterization, especially concerning the cognitive profile, remains a great challenge that can only be achieved through genetic discovery. Indicators show that patients with an X chromosome that is inherited paternally have better verbal skills and interactivity skills (Skuse et al., n.d.). Research on Down syndrome indicates that the 21q22.2 region is a critical region of dosage-sensitive cognition and behaviour-related genes. Research utilizing transgenic mice has revealed a homologue of the Drosophila minibrain gene, identified as a potential gene encoding a tyrosine/serine kinase expressed in the neuroblasts of embryos (Smith et al., 1997; Flint, 2001). This gene is located in the chromosomal region 21q22 and is significant in relation to Down syndrome. Despite the presence of several genes within the Down syndrome critical region (DSCR), there remains a lack of definitive evidence linking these genes to intellectual disability.

2.5 Conclusion

Overall, genetic and chromosomal syndromes are also influential in causing various disabilities, as they interfere with the normal processes of biological development and functions. Genetic and chromosome changes may result in physical, intellectual, developmental, and functional limitations that remain in the life of an individual. Understanding the various types of genetic mutations and chromosomal defects is essential for diagnosing and managing these diseases as early as possible. The significance of inheritance patterns can assist in elucidating the distinctions in the inheritance of genetic disorders among generations and the significance of genetic counselling in averting, educating, and making a wise choice for the family. In addition to medical and biological factors, patients with genetic and chromosomal syndromes tend to encounter significant social difficulties, such as stigmatisation, inaccessible resources, and inclusion barriers. Families and caregivers, who often face emotional strains and economic issues when delivering long-term care and support, also bear these challenges. Thus, it is necessary to focus on a holistic approach that incorporates medical treatment, genetic counselling, social support, and inclusive social policies. Increasing the awareness and knowledge of genetic and chromosomal syndromes may lead to the improvement of the quality of life, support systems, and social acceptance of an affected person and their families.

References

Antonarakis, S. E. (1998). 10 years of genomics, chromosome 21, and down syndrome. *Genomics, 5*(1), 1–6.

Antonarakis, S. E. (2022). Short arms of human acrocentric chromosomes and the completion of the human genome sequence. *Genome Research, 32*(4), 599–607.

Arghir, A., Popescu, R., Resmerita, I., Budisteanu, M., Butnariu, L. I., Gorduza, E. V., Gramescu, M., Panzaru, M. C., Papuc, S. M., Sireteanu, A., & Tutulan-Cunita, A. (2021). Pallister–Killian syndrome versus trisomy 12p—a clinical study of 5 new cases and a literature review. *Genes, 12*(6), 811.

Blackburn, J., Ramakrishnan, A., Graham, C., Bambang, K., Sriranglingam, U., & Senniappan, S. (2025). Klinefelter syndrome: A review. *Clinical Endocrinology, 102*(5), 565–573.

Bossuyt, S. N., Punt, A. M., De Graaf, I. J., Van Den Burg, J., Williams, M. G., Heussler, H., Elgersma, Y., & Distel, B. (2021). Loss of nuclear UBE3A activity is the predominant cause of Angelman syndrome in individuals carrying UBE3A missense mutations. *Human Molecular Genetics, 30*(6), 430–442.

Brand, B. A., Blesson, A. E., & Smith-Hicks, C. L. (2021). The impact of X-chromosome inactivation on phenotypic expression of X-linked neurodevelopmental disorders. *Brain Sciences, 11*(7), 904.

Bufton, J. C., Powers, K. T., Szeto, J. Y., Toelzer, C., Berger, I., & Schaffitzel, C. (2022). Structures of nonsense-mediated mRNA decay factors UPF3B and UPF3A in complex with UPF2 reveal molecular basis for competitive binding and for neurodevelopmental disorder-causing mutation. *Nucleic Acids Research, 50*(10), 5934–5947.

Cao, Y., Luk, H. M., Zhang, Y., Chau, M. H., Xue, S., Cheng, S. S., Li, A. M., Chong, J. S., Leung, T. Y., Dong, Z., & Choy, K. W. (2022). Investigation of chromosomal structural abnormalities in patients with undiagnosed neurodevelopmental disorders. *Frontiers in Genetics, 13*, 803088.

Cardoso, T. M., & Henriques, S. F. (2024). Williams-Beuren region duplication syndrome: Case report. *Arquivos de Neuro-Psiquiatria, 82*(S 02), A003.

Carey, J. C. (2021). Trisomy 18 and trisomy 13 syndromes. *Cassidy and Allanson's Management of Genetic Syndromes*, 937–956.

Červenák, F., Sepšiová, R., Nosek, J., & Tomáška, Ľ. (2021). Step-by-step evolution of telomeres: Lessons from yeasts. *Genome Biology and Evolution, 13*(2), evaa268.

Chen, Y., Dawes, R., Kim, H. C., Ljungdahl, A., Stenton, S. L., Walker, S., Lord, J., Lemire, G., Martin-Geary, A. C., Ganesh, V. S., & Ma, J. (2024). De novo variants in the RNU4-2 snRNA cause a frequent neurodevelopmental syndrome. *Nature, 632*(8026), 832–840.

Chiurazzi, P., Hamel, B. C., & Neri, G. (2001). XLMR genes: Update 2000. *European Journal of Human Genetics, 9*(2), 71–81.

Collins, B. E., & Neul, J. L. (2022). Rett syndrome and MECP2 duplication syndrome: Disorders of MeCP2 dosage. *Neuropsychiatric Disease and Treatment*, 2813–2835.

Cortés-Martín, J., Peñuela, N. L., Sánchez-García, J. C., Montiel-Troya, M., Díaz-Rodríguez, L., & Rodríguez-Blanque, R. (2022). Deletion syndrome 22q11. 2: A systematic review. *Children, 9*(8), 1168.

Cui, C., & Li, P. (2022). FISH—in routine diagnostic settings. In *Cytogenetics and molecular cytogenetics* (pp. 49–70). CRC Press.

Di Stazio, M., Bigoni, S., Iuso, N., Vuch, J., Selvatici, R., Ulivi, S., & d'Adamo, P. A. (2021). Identification of a new mutation in RSK2, the gene for Coffin–Lowry syndrome (CLS), in two related patients with mild and atypical phenotypes. *Brain Sciences, 11*(8), 1105.

Ding, M., Yin, B., Liu, Y., Yao, J., & Dong, H. (2025). Paroxysmal Kinesigenic Dyskinesia with infantile convulsions in a child with inherited co-deletion of 16p11. 2 and 16p12. 2: A case report and literature review. *Cureus, 17*(7).

Doig, K. D., Fellowes, A. P., & Fox, S. B. (2023). Homologous recombination repair deficiency: An overview for pathologists. *Modern Pathology, 36*(3), 100049.

Duis, J., Nespeca, M., Summers, J., Bird, L., Bindels-de Heus, K. G., Valstar, M. J., de Wit, M. C., Navis, C., ten Hooven-Radstaake, M., van Iperen-Kolk, B. M., & Ernst, S. (2022). A multidisciplinary approach and consensus statement to establish standards of care for Angelman syndrome. *Molecular Genetics & Genomic Medicine, 10*(3), e1843.

Efimova, I., Mukhina, A., Markova, Z., Mordanov, S., Soprunova, I., Pershin, D., Balinova, N., Petrusenko, Y., Meleshko, D., Zinchenko, R., & Shilova, N. (2025). 13q deletion syndrome presenting with Lymphopenia detected through newborn screening for primary immunodeficiencies. *International Journal of Molecular Sciences, 26*(19), 9302.

Ferrari, M., & Stagi, S. (2021). Oxidative stress in down and Williams-Beuren syndromes: An overview. *Molecules, 26*(11), 3139.

Fisher, E. M., Beer-Romero, P., Brown, L. G., Ridley, A., McNeil, J. A., Lawrence, J. B., Willard, H. F., Bieber, F. R., & Page, D. C. (1990). Homologous ribosomal protein genes on the human X and Y chromosomes: Escape from X inactivation and possible implications for Turner syndrome. *Cell, 63*(6), 1205–1218.

Flint, J. (2001). Genetic basis of cognitive disability. *Dialogues in Clinical Neuroscience, 3*(1), 37–46.

Frayling, T. M., Timpson, N. J., Weedon, M. N., Zeggini, E., Freathy, R. M., Lindgren, C. M., Perry, J. R., Elliott, K. S., Lango, H., Rayner, N. W., & Shields, B. (2007). A common variant in the FTO gene is associated with body mass index and predisposes to childhood and adult obesity. *Science, 316*(5826), 889–894.

Garshasbi, M., Hadavi, V., Habibi, H., Kahrizi, K., Kariminejad, R., Behjati, F., Tzschach, A., Najmabadi, H., Ropers, H. H., & Kuss, A. W. (2008). A defect in the TUSC3 gene is associated with autosomal recessive mental retardation. *The American Journal of Human Genetics, 82*(5), 1158–1164.

Gavril, E. C., Luca, A. C., Curpan, A. S., Popescu, R., Resmerita, I., Panzaru, M. C., Butnariu, L. I., Gorduza, E. V., Gramescu, M., & Rusu, C. (2021). Wolf-Hirschhorn syndrome: Clinical and genetic study of 7 new cases, and mini review. *Children, 8*(9), 751.

Gawade, K., & Raczynska, K. D. (2024). Imprinted small nucleolar RNAs: Missing link in development and disease? *Wiley Interdisciplinary Reviews: RNA, 15*(1), e1818.

Girirajan, S., Rosenfeld, J. A., Cooper, G. M., Antonacci, F., Siswara, P., Itsara, A., Vives, L., Walsh, T., McCarthy, S. E., Baker, C., & Mefford, H. C. (2010). A recurrent 16p12. 1 microdeletion supports a two-hit model for severe developmental delay. *Nature Genetics, 42*(3), 203–209.

Good, K. V., Vincent, J. B., & Ausió, J. (2021). MeCP2: The genetic driver of Rett syndrome epigenetics. *Frontiers in Genetics, 12*, 620859.

Gravholt, C. H., Viuff, M., Just, J., Sandahl, K., Brun, S., van der Velden, J., Andersen, N. H., & Skakkebaek, A. (2023). The changing face of Turner syndrome. *Endocrine Reviews, 44*(1), 33–69.

Hamad, A., Sherlaw-Sturrock, C. A., Glover, K., Salmon, R., Low, K., Nair, R., Sansbury, F. H., Rawlins, L., Carmichael, J., Horton, R., & Wedderburn, S. (2023). Expanding the phenotypic spectrum of Chromosome 16p13. 11 microduplication: A multicentric analysis of 206 patients. *European Journal of Medical Genetics, 66*(4), 104714.

Higgins, J. J., Pucilowska, J., Lombardi, R. Q., & Rooney, J. P. (2004). A mutation in a novel ATP-dependent Lon protease gene in a kindred with mild mental retardation. *Neurology, 63*(10), 1927–1931.

Hirvikoski, T., Boman, M., Tideman, M., Lichtenstein, P., & Butwicka, A. (2021). Association of intellectual disability with all-cause and cause-specific mortality in Sweden. *JAMA Network Open, 4*(6), e2113014.

Huisman, S., Festen, D., & Bakker-van, G. E. (2024). Healthcare for people with intellectual disabilities in The Netherlands. *Journal of Policy and Practice in Intellectual Disabilities, 21*(2), e12496.

Jacquin, C., Landais, E., Poirsier, C., Afenjar, A., Akhavi, A., Bednarek, N., Bénech, C., Bonnard, A., Bosquet, D., Burglen, L., & Callier, P. (2023). 1p36 deletion syndrome: Review and mapping with further characterization of the phenotype, a new cohort of 86 patients. *American Journal of Medical Genetics, Part A, 191*(2), 445–458.

Jansen, S., Vissers, L. E., & de Vries, B. B. (2023). The genetics of intellectual disability. *Brain Sciences, 13*(2), 231.

Jin, P., & Warren, S. T. (2000). Understanding the molecular basis of fragile X syndrome. *Human Molecular Genetics, 9*(6), 901–908.

Kaya, M., Suer, I., Kalayci, T., Karaman, B., Ozturk, S., & Palanduz, S. (2022). Cytogenetic and molecular characterization of a patient having infertility and mild intellectual disability with a very rare unstable ring chromosome 13. *Scottish Medical Journal, 67*(4), 173–177.

Kikas, T., Dutta, A., Inno, R., Pomm, K., Tjagur, S., Poolamets, O., Roomere, H., Punab, M., & Laan, M. (2025). Microdeletion and microduplication syndromes, including recurrent rearrangements at 16p11. 2 and 22q11. 21, are enriched in unexplained male infertility. *Human Reproduction*, deaf231.

Kolbasin, L. N., Dubrovskaya, T. A., Salnikova, G. B., Solovieva, E. N., Donnikov, M. Y., Illarionov, R. A., Glotov, A. S., Kovalenko, L. V., & Belotserkovtseva, L. D. (2024). Family case of Potocki-Lupski syndrome. *Molecular Cytogenetics, 17*(1), 6.

Kozel, B. A., Boaz, B., Kim, C. A., Mervis, C. B., Osborne, L. R., Porter, M., & Pober, B. R. (2021). Williams syndrome (primer). *Nature Reviews: Disease Primers, 7*(1).

Kumar, A., Abbas, A. K., & Jon, C. J. (2015). Aster: Robbins and Cotran pathologic basis of disease. *Professional Edition*.

Kuss, A. W., Garshasbi, M., Kahrizi, K., Tzschach, A., Behjati, F., Darvish, H., Abbasi-Moheb, L., Puettmann, L., Zecha, A., Weißmann, R., & Hu, H. (2011). Autosomal recessive mental retardation: Homozygosity mapping identifies 27 single linkage intervals, at least 14 novel loci and several mutation hotspots. *Human Genetics, 129*(2), 141–148.

Laxova, R., Ridler, M. A., Bowen-Bravery, M., & Opitz, J. M. (1977). An etiological survey of the severely retarded Hertfordshire children who were born between January 1, 1965 and December 31, 1967. *American Journal of Medical Genetics., 1*(1), 75–86.

Lewis, B. A. (2021). Familial and genetic bases of speech and language disorders 1. *Attention, Genes and ADHD*, 80–98.

Lindstrand, A., Ek, M., Kvarnung, M., Anderlid, B. M., Bjoerck, E., Carlsten, J., Eisfeldt, J., Grigelioniene, G., Gustavsson, P., Hammarsjoe, A., & Helgadottir, H. T. (2022). Genome sequencing is a sensitive first-line test to diagnose individuals with intellectual disability. *Genetics in Medicine, 24*(11), 2296–2307.

Liu, S., Chen, M., Yang, H., Chen, S., Wang, L., Duan, L., Zhu, H., & Pan, H. (2021). Clinical characteristics and long-term recombinant human growth hormone treatment of 18q-syndrome: A case report and literature review. *Frontiers in Endocrinology, 12*, 776835.

Lovras, M., Rajput, S., Sridhar, S. B., Shareef, J., & Malviya, R. (2025). Cancer management using photodynamic therapy: Fundamentals, mechanism and advances. *Current Pharmaceutical Design*.

Lowther, C., Mehrjouy, M. M., Collins, R. L., Bak, M. C., Dudchenko, O., Brand, H., Dong, Z., Rasmussen, M. B., Gu, H., Weisz, D., & Nazaryan-Petersen, L. (2022). Balanced chromosomal rearrangements offer insights into coding and noncoding genomic features associated with developmental disorders. *medrxiv*, 2022.

Lühmann, J. L., Schmidt, G., Auber, B., Bergmann, A. K., Brandau, O., Louis, A., Hentze, S., Eisfeld, K., Schlegelberger, B., Klaes, R., & Steinemann, D. (2023). Parallel deletion and duplication at 7q11. 23 in a silent carrier for two reciprocal syndromic disorders. *American Journal of Medical Genetics Part A, 191*(7), 1849–1857.

Lukhtanov, V. A. (2022). Diversity and evolution of telomere and subtelomere DNA sequences in insects. *BioRxiv*, 2022.

Luthardt, F. W., & Keitges, E. Chromosomal syndromes and genetic disease. e LS. 2001 May 30.

Malviya, R., & Rajput, S. (2025). *Artificial intelligence for neural health: Diagnosis and treatment*. CRC Press.

Manning, L. K., Winkenwerder, E., Baskind, L., Eager, K. L., Willet, C. E., Porebski, B., O'Rourke, B. A., Tammen, I., Gimeno, M., & Pinczowski, P. (2025). A novel missense variant in the RELN gene in sheep with lissencephaly and cerebellar hypoplasia. *Veterinary Pathology*, 03009858241283501.

Matasariu, D. R., Bujor, I. E., Gireada, R. M., Guzga, L. M., Nedelea, F. M., Titianu, M., & Ursache, A. (2025). Challenges in prenatal ultrasound diagnosis of Rubinstein–Taybi syndrome: A case report and comprehensive literature review. *International Journal of Molecular Sciences, 26*(11), 5142.

McDonald-McGinn, D. M., Jeong, S., McGinn, M. J., Zackai, E. H., & Unolt, M. (2021). Deletion 22q11. 2 (velo-cardio-facial syndrome/DiGeorge syndrome). *Cassidy and Allanson's Management of Genetic Syndromes*, 291–316.

Meyyazhagan, A., & Di Renzo, G. C. (2023). The nexus between chromosomal abnormalities and single gene disorders. In *Prenatal diagnostic testing for genetic disorders: The revolution of the non-invasive prenatal test* (pp. 25–56). Springer.

Molinari, F., Foulquier, F., Tarpey, P. S., Morelle, W., Boissel, S., Teague, J., Edkins, S., Futreal, P. A., Stratton, M. R., Turner, G., & Matthijs, G. (2008). Oligosaccharyltransferase-subunit mutations in nonsyndromic mental retardation. *The American Journal of Human Genetics, 82*(5), 1150–1157.

Morris, C. A., & Mervis, C. B. (2021). Williams syndrome. *Cassidy and Allanson's Management of Genetic Syndromes*, 1021–1038.

Muscogiuri, G., Barrea, L., Faggiano, F., Maiorino, M. I., Parrillo, M., Pugliese, G., Ruggeri, R. M., Scarano, E., Savastano, S., Colao, A., & RESTARE. (2021). Obesity in Prader–Willi syndrome: Physiopathological mechanisms, nutritional and pharmacological approaches. *Journal of Endocrinological Investigation, 44*(10), 2057–2070.

Najmabadi, H., Motazacker, M. M., Garshasbi, M., Kahrizi, K., Tzschach, A., Chen, W., Behjati, F., Hadavi, V., Nieh, S. E., Abedini, S. S., & Vazifehmand, R. (2007). Homozygosity mapping in consanguineous families reveals extreme heterogeneity of non-syndromic autosomal recessive mental retardation and identifies 8 novel gene loci. *Human Genetics, 121*(1), 43–48.

Olsen, J. A., Chen, G., & Lamu, A. N. (2023). The relative importance of education and health behaviour for health and wellbeing. *BMC Public Health, 23*(1), 1981.

Olusanya, B. O., Gladstone, M., Wright, S. M., Hadders-Algra, M., Boo, N. Y., Nair, M. K., Almasri, N., Kancherla, V., Samms-Vaughan, M. E., Kakooza-Mwesige, A., & Smythe, T. (2022). Cerebral palsy and developmental intellectual disability in children younger than 5 years: Findings from the GBD-WHO rehabilitation database 2019. *Frontiers in Public Health, 10*, 894546.

Papadopoulou, S., Anagnostopoulou, A., Katsarou, D. V., Megari, K., Efthymiou, E., Argyriadis, A., Kougioumtzis, G., Theodoratou, M., Sofologi, M., Argyriadi, A., & Pavlidou, E. (2024). Enhancing communication and swallowing skills in children with cri Du chat syndrome: A comprehensive speech therapy guide. *Children, 11*(12), 1526.

Papenhausen, P. R., Kelly, C. A., Harris, S., Caldwell, S., Schwartz, S., & Penton, A. (2021). Clinical significance and mechanisms associated with segmental UPD. *Molecular Cytogenetics, 14*(1), 38.

Pascual, U., Balvanera, P., Anderson, C. B., Chaplin-Kramer, R., Christie, M., González-Jiménez, D., Martin, A., Raymond, C. M., Termansen, M., Vatn, A., & Athayde, S. (2023). Diverse values of nature for sustainability. *Nature, 620*(7975), 813–823.

Pascual-Alonso, A., Martínez-Monseny, A. F., Xiol, C., & Armstrong, J. (2021). MECP2-related disorders in males. *International Journal of Molecular Sciences, 22*(17), 9610.

Patil, S., Rao, R. S., & Majumdar, B. (2014). Chromosomal and multifactorial genetic disorders with oral manifestations. *Journal of International Oral Health, 6*(5), 118.

Phelan, K., Rogers, R. C., & Boccuto, L. (2021). Deletion 22q13 syndrome: Phelan–McDermid syndrome. *Cassidy and Allanson's Management of Genetic Syndromes*, 317–334.

Piao, J., Huang, Y., Han, C., Li, Y., Xu, Y., Liu, Y., & He, X. (2022). Alarming changes in the global burden of mental disorders in children and adolescents from 1990 to 2019: A systematic analysis for the global burden of disease study. *European Child & Adolescent Psychiatry, 31*(11), 1827–1845.

Pisano, M., Sangiovanni, G., D'Ambrosio, F., Romano, A., & Di Spirito, F. (2022). Oral care in a patient with long arm deletion syndrome of chromosome 18: A narrative review and case presentation. *The American Journal of Case Reports, 23*, e936142.

Pomohaibo, V., Karapuzova, N., & Pavlenko, Y. (2021). Finding genetic factors associated with cognitive abilities. *ScienceRise: Pedagogical Education, 6*(45), 29–34.

Purow, J., Waidner, L., & Ale, H. (2025). Review of the pathophysiology and clinical manifestations of 22q11. 2 deletion and duplication syndromes. *Clinical Reviews in Allergy & Immunology, 68*(1), 23.

Rajput, S., Malviya, R., Bahadur, S., & Puri, D. (2023). Recent updates on the development of therapeutics for the targeted treatment of Alzheimer's disease. *Current Pharmaceutical Design., 29*(35), 2802–2813.

Rajput, S., Malviya, R., & Uniyal, P. (2024). Advances in the treatment of kidney disorders using mesenchymal stem cells. *Current Pharmaceutical Design, 30*(11), 825–840.

Rivera, M., Solari, G., & Peralta, M. (2021). Estudio de prevalencia en niños recién nacidos con síndrome Down y sus características antropométricas. Hospital Regional de Antofagasta, Chile. *Revista chilena de nutrición, 48*(2), 238–244.

Rogers, R. C. (2021). Coffin–Lowry syndrome. *Cassidy and Allanson's Management of Genetic Syndromes, 171*, –84.

Saia, F., Prato, A., Florio, C. A., Cutrone, V. P., & Rizzo, R. (2023). 17q21. 31 microduplication syndrome in a patient with autism spectrum disorder, macrocephaly, and intellectual disability. *Report, 6*(3), 30.

Samuel, D., O'Malley, F., Brink, F. W., Crichton, K. G., Duffy, B., Letson, M. M., & Michaels, N. L. (2023). Characterizing child maltreatment fatalities among child victims with disabilities in the United States, 2010–2019. *Child Abuse & Neglect, 144*, 106354.

Schmidt, P., Suchy, L. C., & Schulte, A. G. (2022). Oral health care of people with down syndrome in Germany. *International Journal of Environmental Research and Public Health, 19*(19), 12435.

Schuy, J., Grochowski, C. M., Carvalho, C. M., & Lindstrand, A. (2022). Complex genomic rearrangements: An underestimated cause of rare diseases. *Trends in Genetics, 38*(11), 1134–1146.

Scott, W. A., Dhanji, E. Z., Dyakov, B. J., Dreseris, E. S., Asa, J. S., Grange, L. J., Mirceta, M., Pearson, C. E., Stewart, G. S., Gingras, A. C., & Campos, E. I. (2021). ATRX proximal protein associations boast roles beyond histone deposition. *PLoS Genetics, 17*(11), e1009909.

Shaffer, L. G. (2021). Special issue on companion animal genetics: Novel variants discovered in wide variety of diseases in dogs, identification and further characterization of traits in dogs and cats, and the use of microarrays in the detection of aneuploidy in dogs. *Human Genetics, 140*(11), 1501–1503.

Shaffer, L. G., & Lupski, J. R. (2000). Molecular mechanisms for constitutional chromosomal rearrangements in humans. *Annual Review of Genetics, 34*(1), 297–329.

Skuse, D. H., James, R. S., Bishop, D. V., Coppin, B., Dalton, P., Aamodt-Leeper, G., Bacarese-Hamilton, M., Creswell, C., McGurk, R., & Jacobs, P. A.. (n.d.) Evidence from Turner's syndrome of an imprinted X-linked locus affecting cognitive function.

Smith, D. J., Stevens, M. E., Sudanagunta, S. P., Bronson, R. T., Makhinson, M., Watabe, A. M., O'Dell, T. J., Fung, J., Weier, H. U., Cheng, J. F., & Rubin, E. M. (1997). Functional screening of 2 Mb of human chromosome 21q22. 2 in transgenic mice implicates minibrain in learning defects associated with down syndrome. *Nature Genetics, 16*(1), 28–36.

Song, C., & Broadie, K. (2023). Fragile X mental retardation protein coordinates neuron-to-glia communication for clearance of developmentally transient brain neurons. *National Academy of Sciences of the United States of America, 120*(12), e2216887120.

Stolz, J. R., Foote, K. M., Veenstra-Knol, H. E., Pfundt, R., Ten Broeke, S. W., de Leeuw, N., Roht, L., Pajusalu, S., Part, R., Rebane, I., & Õunap, K. (2021). Clustered mutations in the GRIK2 kainate receptor subunit gene underlie diverse neurodevelopmental disorders. *The American Journal of Human Genetics, 108*(9), 1692–1709.

Sun, K. Y., Bai, X., Chen, S., Bao, S., Zhang, C., Kapoor, M., Backman, J., Joseph, T., Maxwell, E., Mitra, G., & Gorovits, A. (2024). A deep catalogue of protein-coding variation in 983,578 individuals. *Nature, 631*(8021), 583–592.

Swierkowska-Janc, J., Kabza, M., Rydzanicz, M., Giefing, M., Ploski, R., Shaffer, L. G., & Gajecka, M. (2025). DNA methylation dysregulation patterns in the 1p36 region instability. *Journal of Applied Genetics, 66*(3), 611–621.

Szczawińska-Popłonyk, A., Schwartzmann, E., Chmara, Z., Głukowska, A., Krysa, T., Majchrzycki, M., Olejnicki, M., Ostrowska, P., & Babik, J. (2023). Chromosome 22q11. 2 deletion syndrome: A comprehensive review of molecular genetics in the context of multidisciplinary clinical approach. *International Journal of Molecular Sciences, 24*(9), 8317.

Talebizadeh, Z., Simon, S. D., & Butler, M. G. (2006). X chromosome gene expression in human tissues: Male and female comparisons. *Genomics, 88*(6), 675–681.

Talon, I., Janiszewski, A., Theeuwes, B., Lefevre, T., Song, J., Bervoets, G., Vanheer, L., De Geest, N., Poovathingal, S., Allsop, R., & Marine, J. C. (2021). Enhanced chromatin accessibility contributes to X chromosome dosage compensation in mammals. *Genome Biology, 22*(1), 302.

Theisen, A., & Shaffer, L. G. (2010). Disorders caused by chromosome abnormalities. *The Application of Clinical Genetics*, 159–174.

Tolmacheva, E. N., Fonova, E. A., & Lebedev, I. N. (2022). X-linked CNV in pathogenetics of intellectual disability. *Russian Journal of Genetics, 58*(10), 1193–1207.

Townsend, A. K., Williams, K. E., & Nannas, N. J. (2025). Inbreeding and cognitive impairment in animals. *Behavioral Ecology, 36*(1), arae101.

Trevisan, V., Meroni, A., Leoni, C., Sirchia, F., Politano, D., Fiandrino, G., Giorgio, V., Rigante, D., Limongelli, D., Perri, L., & Sforza, E. (2024). Trisomy 22 mosaicism from prenatal to postnatal findings: A case series and systematic review of the literature. *Genes, 15*(3), 346.

Van Gils, J., Magdinier, F., Fergelot, P., & Lacombe, D. (2021). Rubinstein-Taybi syndrome: A model of epigenetic disorder. *Genes, 12*(7), 968.

Vervoort, L., & Vermeesch, J. R. (2022). The 22q11. 2 low copy repeats. *Genes, 13*(11), 2101.

Vitale, G., Mattiaccio, A., Conti, A., Turco, L., Seri, M., Piscaglia, F., & Morelli, M. C. (2022). Genetics in familial intrahepatic cholestasis: Clinical patterns and development of liver and biliary cancers: A review of the literature. *Cancers, 14*(14), 3421.

Vollger, M. R., Guitart, X., Dishuck, P. C., Mercuri, L., Harvey, W. T., Gershman, A., Diekhans, M., Sulovari, A., Munson, K. M., Lewis, A. P., & Hoekzema, K. (2022). Segmental duplications and their variation in a complete human genome. *Science, 376*(6588), eabj6965.

Wang, Y., Ma, Q., Chen, J., Li, S., Zheng, F., Shi, L., Li, X., Li, S., Tong, G., & Li, H. (2024). Identification of a Novel Frameshift variant of the ATRX gene: A case report and review of the genotype–phenotype relationship. *BMC Pediatrics, 24*(1), 631.

Wang, G., Xu, Y., Wang, Q., Chai, Y., Sun, X., Yang, F., Zhang, J., Wu, M., Liao, X., Yu, X., & Sheng, X. (2022). Rare and undiagnosed diseases: From disease-causing gene identification to mechanism elucidation. *Fundamental Research, 2*(6), 918–928.

Xi, H., Ma, L., Yin, X., Yang, P., Li, X., & Li, L. (2024). X-linked intellectual developmental disorder with onset of neonatal heart failure: A case report and literature review. *Molecular Genetics and Metabolism Reports, 38*, 101054.

Yang, J., Chen, Z., He, J., Zou, B., Si, Y., Ma, Y., & Yu, J. (2025). Modular RNA interactions shape FXR1 condensates involved in mRNA localization and translation. *Nature Communications, 16*(1), 8589.

Yong, Z., Dou, Y., Gao, Y., Xu, X., Xiao, Y., Zhu, H., Li, S., & Yuan, B. (2021). Prenatal, perinatal, and postnatal factors associated with autism spectrum disorder cases in Xuzhou, China. *Translational Pediatrics, 10*(3), 635.

Zedan, H. S., Bilal, L., Hyder, S., Naseem, M. T., Akkad, M., Al-Habeeb, A., Al-Subaie, A. S., & Altwaijri, Y. (2023). Understanding the burden of mental and physical health disorders on families: Findings from the Saudi National Mental Health Survey. *BMJ Open, 13*(10), e072115.

Zhang, X., Fan, J., Chen, Y., Wang, J., Song, Z., Zhao, J., Li, Z., Wu, X., & Hu, Y. (2021). Cytogenetic analysis of the products of conception after spontaneous abortion in the first trimester. *Cytogenetic and Genome Research., 161*(3–4), 120–131.

Zhang, Y., Liu, X., Gao, H., He, R., Chu, G., & Zhao, Y. (2021). Copy number variations of chromosome 17p11. 2 region in children with development delay and in fetuses with abnormal imaging findings. *BMC Medical Genomics, 14*(1), 215.

Zhang, K., Liu, S., Gu, W., Lv, Y., Yu, H., Gao, M., Wang, D., Zhao, J., Li, X., Gai, Z., & Zhao, S. (2021). Transmission of a novel imprinting center deletion associated with Prader–Willi syndrome through three generations of a chinese family: Case presentation, differential diagnosis, and a lesson worth thinking about. *Frontiers in Genetics, 12*, 630650.

Zhou, Y., Wang, Y., Xiong, X., Appel, A. G., Zhang, C., & Wang, X. (2022). Profiles of telomeric repeats in Insecta reveal diverse forms of telomeric motifs in hymenopterans. *Life Science Alliance, 5*(7).

Zhou, J., Zheng, Y., Liang, G., Xu, X., Liu, J., Chen, S., Ge, T., Wen, P., Zhang, Y., Liu, X., & Zhuang, J. (2022). Atypical deletion of Williams–Beuren syndrome reveals the mechanism of neurodevelopmental disorders. *BMC Medical Genomics, 15*(1), 79.

Zoyirov, T. E., & Indiaminova, G. N. (2021). Improvement of methods of providing dental care for children with mental delay. *Central Asian Journal of Medical and Natural Science, 2*(6), 167–170.

Chapter 3
Nervous System Disorders and Disability

Abstract The nervous system controls and regulates all bodily functions while transmitting signals from the brain, spinal cord, and various organs throughout the body. This chapter provides an in-depth analysis of neurological disorders associated with the nervous system that cause a disability. The chapter explains the way injuries or dysfunctions to the spinal cord, brain, and peripheral nerves can disrupt movement, sensation, cognition, and behaviour. Furthermore, the chapter illustrates various clinical manifestations and the complications of the involvement of the nervous system in disability. Additionally, the chapter explains the way people with nervous system disorders and disabilities (epilepsy, cerebral palsy, multiple sclerosis, etc.) frequently experience social difficulties such as movement impairments, inaccessibility to education and jobs, and social exclusion, which has a profound impact on both the affected person and his or her family.

Keywords Nervous system disorders · Neurological disability · Brain and spinal cord dysfunction · Motor impairment · Cognitive impairment · Epilepsy · Multiple sclerosis · Parkinson's disease

3.1 Introduction

The World Health Organization defines disability as "any limitation or inability to execute an activity deemed normal for a human being as a result of an impairment" (Prabhu et al., 2025). Disability may be evaluated by both subjective and objective data. Inquiries concerning the performance of essential activities of daily living (ADLs), including dressing and bathing, alongside instrumental activities of daily living (IADLs) such as shopping and meal preparation, provide valuable insights into the patient's understanding of their limitations. During the office visit, objective data concerning impairment is often assessed, as evidenced by the patient's difficulty in undressing for the examination or managing toileting independently. It is possible to obtain objective, quantitative data on functional abilities through

R. Malviya, S. Rajput, *Neurogenetic and Neurodevelopmental Disabilities*, SpringerBriefs in Modern Perspectives on Disability Research,
https://doi.org/10.1007/978-981-95-9223-4_3

physical performance testing, which includes timed gait tests and daily activity simulations. The major epidemiological research points to the fact that neurological diseases are significant causes of mortality and disability on a global scale, affecting one out of nine people. Among these conditions, stroke is the commonest cause of death. Several problems arise in poor countries, such as developmental deviations, which are a result of starvation and cognitive retardation brought about by parasitic diseases. The number of disorders like Parkinson's, Alzheimer's, and epilepsy are on the rise as the world population gets older, and infectious diseases are on the decrease (World Health Organization, 2023). Most of these disorders are preventable and treatable, which explains the need for the medical sector in the neurological field to deserve increased awareness and focus in health care planning, research, and administration.

About one in nine people die of nervous system diseases that represent about 28 years of disability life (Murray, 2022). Cerebrovascular disease ranks as the second foremost cause of mortality worldwide, exhibiting a notably elevated prevalence compared to other neurological disorders. The most widespread neurological disease is major depression, and alcoholism comes second (Murray, 2024). The aging demographic, particularly in the developing countries, will likely cause more prevalence of stroke, dementia, and related diseases; hence, more mortality and disability due to these diseases (Bergen & Silberberg, 2002). The nervous system is a complex network that is important in the process of the regulation and coordination of various physiological functions. The structure is divided into two main parts: the central nervous system (CNS) and the peripheral nervous system (PNS) (Banerjee et al., 2023). Information processing occurs at the central nervous system, which is represented by the brain and the spinal cord. The peripheral nervous system comprises peripheral and autonomic nerves, which are distributed in every part of the body, and they facilitate the transmission of information between the central nervous system and the rest of the organs in the organism. The main elements of the nervous system include the brain, which processes sensory information, and the spinal cord, which facilitates the communication of the brain with the peripheral nerves. Moreover, the sense organs like eyes and ears are important in vision and hearing, respectively (Park, 2023). The system contains skin, joint, and muscle sensory receptors, as well as sensitivity organs (taste and smell) that monitor ambient stimuli and physical position or movement. The nervous system plays an important role in maintaining homeostasis and interactions with the immediate environment (Myers Jr et al., 2021).

Many diseases may develop in the nervous system, causing a state of impairment. These encompass physical damage to neural structures, infections affecting the central or peripheral nervous systems, and degenerative diseases that gradually impair brain functions (Teixeira, 2024). Additionally, anomalies of the brain development may be hindered by congenital anatomical anomalies. The tumours can press on the neural tissues, thus inducing impairment of the functions, and also the alterations of blood flow may prevent the supply of essential nutrients to the brain and nerve cells. In addition, autoimmune diseases can act improperly and target brain tissue, thus causing inflammation and damage. All of these factors prove the complexity of the treatment of the nervous system disorders (Xiang et al., 2023). Neurodisability may be caused by various neurological diseases, all of which are categorized on the basis of their etiological causes. Vascular disorders encompass a range of conditions, including stroke, subarachnoid haemorrhage, hematoma, subdural haemorrhage, transient ischemic attack (TIA), and extradural haemorrhage (Kurz & Rogers, 2022). The infections affecting the nervous system encompass meningitis, encephalitis, myelitis, and epidural abscess (Velnar et al., 2023). Structural abnormalities may arise from various conditions, including brain or spinal cord injuries, carpal tunnel syndrome, cervical spondylosis, and tumours affecting the brain or spinal cord (Saifee et al., 2024). The other group of disorders that disrupt the normal functioning of the nervous system are seizure disorders, with epilepsy being the primary example. The progressive disorders with progressive disability are Parkinson's, Huntington's, amyotrophic lateral sclerosis (ALS), and Alzheimer's. In addition, other autoimmune and inflammatory diseases like Bell's palsy, peripheral neuropathy, multiple sclerosis, and Guillain-Barré syndrome may impact nervous system.

Neurodisability is also complicated by the occurrence of mental illnesses, such as mood disorder, depression, and schizophrenia. These circumstances highlight the numerous variables that may deter the nervous system, as well as the subjects of the disorders (Garzon & Starr, 2023). The dysfunction of the nervous system can manifest itself in diverse ways in different individuals and, in the majority of cases, can be described as several neurological symptoms. The most frequent ones are the recurrent headaches or the sudden ones, which may be of different nature or intensity. Sensory loss or paraesthesia, muscular weakness or atrophy, and visual impairment, such as blindness or diplopia, may occur in people. Moreover, possible indicative factors of dysfunction may be cognitive deficiency and impaired memory. The coordination issues are prevalent, and such symptoms as muscle rigidity, seizures, and tremors are observed (Bravo et al., 2023). The symptom of back pain in the feet, toes, or other parts of the body can be associated with the conditions of the nervous system, which can subsequently lead to atrophy of muscles and impeding of speech. Moreover, language impairments may arise, and they may be expressive or comprehensive. It is also important to note that the symptoms might seem similar to the symptoms of other diseases; hence, it is prudent to seek medical attention from someone who might understand the symptoms appropriately. Figure 3.1 and Table 3.1 illustrates various nervous system disorders and disability.

3.2 Various Nervous System Disorders and Disabilities

3.2.1 Alzheimer's Disease

Dementia is a significant impairment of the cognitive system that affects everyday life. The predominant form of dementia is known as Alzheimer's disease (AD), which is identified as occurring in at least two out of every three people who have

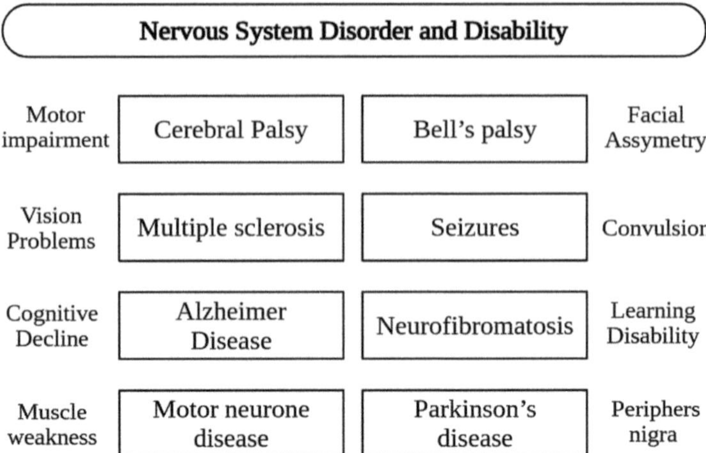

Fig. 3.1 Nervous system disorders and disability

reached the age of 65 or older. Alzheimer's disease refers to a condition characterized by a gradual decline of cognitive and behavioural ability and serves as a primary factor of death (Ahmad, 2021). It usually starts after age 65 (late-onset Alzheimer's disease), though it can start earlier (early-onset Alzheimer's disease) in about 5% of cases, often with non-characteristic symptoms and accelerating progression (Rajput et al., 2023). The latest discoveries include early diagnosis signs, such as neuroimaging and measurement of amyloid and tau in cerebrospinal fluid and plasma (Zetterberg & Bendlin, 2021). There exists currently no treatment for Alzheimer's disease (AD), though therapies can be used to treat the symptoms. Among the new findings, there is the introduction of novel pharmacological agents aimed at disease progression, as well as the discovery of new biomarkers (Porsteinsson et al., 2021). Alzheimer's disease appears to develop in preclinical and severe stages, characterized by varying degrees of cognitive dysfunction. The most common early symptom is the intermittent short-term memory deficits, which progress to executive functioning deficits, which affect day-to-day functioning, including driving and money management (Therriault et al., 2022). As the disease progresses, patients may develop language deficiencies, dysfunctions in visuospatial functioning, and other neuropsychiatric symptoms. At the progressive stages, difficulties in motor functions, sleep disturbances, and complete dependency on caretakers appear (Maccioni et al., 2018).

AD represents a progressive neurodegenerative disorder characterized by the degeneration of neuronal cells, initially manifesting in the entorhinal cortex of the hippocampus. Genetic factors impact early and late emergent Alzheimer's diseases, and trisomy 21 represents a significant risk factor for the development of early-onset dementia (Santos et al., 2017). The first risk factor is age, whereby the incidence of Alzheimer's disease approximately doubles every 5 years following the age of 65. The number of cardiovascular diseases (CVD), obesity, and diabetes are significant

Table 3.1 Nervous system disorders and disability

Disorder	Primary affected system	Pathophysiology	Key neurological disabilities
Alzheimer'sdisease	Central nervous system	Progressive neurodegeneration with amyloid-β plaques and tau tangles	Memory loss, cognitive decline, impaired judgment, behavioural changes
Bell's palsy	Peripheral nervous system (facial nerve VII)	Acute inflammation or compression of the facial nerve	Facial muscle weakness, facial asymmetry, speech and eating difficulty
Cerebral palsy	Central nervous system	Non-progressive brain injury during the prenatal or early postnatal period	Motor impairment, muscle spasticity, coordination and posture deficits
Seizures (epilepsy)	Central nervous system	Abnormal, excessive neuronal electrical activity	Loss of consciousness, convulsions, cognitive impairment
Multiple sclerosis (MS)	Central nervous system	Autoimmune-mediated demyelination of neurons	Muscle weakness, sensory loss, visual disturbance, fatigue
Neurofibromatosis	Central and peripheral nervous systems	Genetic mutation causing tumour formation along nerves	Learning disabilities, motor impairment, vision/hearing loss
Parkinson's disease	Central nervous system	Degeneration of dopaminergic neurons in the substantia nigra	Tremor, rigidity, bradykinesia, postural instability
Motor neuron disease (ALS)	Central and peripheral motor neurons	Progressive degeneration of upper and lower motor neurons	Muscle weakness, paralysis, speech and swallowing difficulties

modifiable risk factors. Cardiovascular disease (CVD) predisposes the advancement of AD and dementia (stroke-induced), and obesity and diabetes can negatively affect glucose tolerance and exacerbate the cognitive decline by beta-amyloid accumulation and neuroinflammation (Anjum et al., 2018). Other risk factors of Alzheimer's disease (AD) include traumatic brain damage, depression, heart disease, late age of parental birth, use of tobacco, family history of dementia, high levels of homocysteine, and the existence of the APOE e4 allele. The risk probability runs between 10 and 30% higher in first-degree relatives of AD patients, which may increase three times with multiple siblings affected by the disease (Liljegren et al., 2018). Risk factors that are mitigating are a high level of education, hormone/anti-inflammatory drugs among females, leisure activities, a healthy diet, and regular aerobic exercises. The prevalence of dementia had escalated to 43.8 million cases in 2016, a significant increase from 20.3 million cases recorded in 1990. Projections indicate that this figure may soar to 150 million by the year 2050. The prevalence rate of developing Alzheimer's disease is increasing at an average age of 65 years after every 5 years, and this is particularly true in women above the age of 85 years (Qiu et al., 2009).

3.2.1.1 Pathophysiology Genetic Basis of Alzheimer's Disease

The accumulation of abnormal neurotic plaques and neurofibrillary tangles leads to neuronal degeneration, particularly that of cholinergic neurons, which is typical of Alzheimer's disease (Breijyeh & Karaman, 2020). It can be described by two key ideas: The cholinergic hypothesis suggests that the depletion of acetylcholine (ACh) levels, which is caused by the loss of cholinergic cells, plays a major role in the development of Alzheimer's disease. Moreover, beta-amyloid is also thought to have a negative impact on cholinergic functioning (Hampel et al., 2018). The Amyloid Hypothesis is based on the idea that an increase in the concentration of amyloid beta (Aβ) peptide, which is one of the components of the amyloid precursor protein (APP) formed under the influence of specific enzymes, causes neuronal toxicity and stimulates the development of toxic amyloid aggregates (Paroni et al., 2019). Trisomy 21 is the cause of Down syndrome, and it is closely related to the amyloid input to the study of Alzheimer's disease. People with Down syndrome have a second copy of the APP gene, which causes the overproduction of amyloid-β (Aβ) and the exposure to more risk of developing AD. Alzheimer's disease in the clinical sense is expected to develop in 40–80% of persons with Down syndrome in their fifth or sixth decade of life. Moreover, the signs of Alzheimer's disease pathology, such as amyloid plaques and neurofibrillary tangles, are detected in almost all people with Down syndrome (Salehi et al., 2016). Alzheimer's disease may be an inherent disorder that is transmitted when an individual is affected by an autosomal dominant disorder, which is caused by mutations in three genes, namely APP (located on chromosome 21), Presenilin 1 (PSEN1 (14)), and Presenilin 2 (PSEN2 (1)). Though these mutations are sporadic (about 5–10% of cases of AD), they are linked to an early-onset type of the disease, and PSEN1 is the most prevalent mutation (Hoogmartens et al., 2021). The APOE is part of lipid metabolism that is very relevant in relation to the beta-amyloid protein. There exist three distinct alleles, specifically ε2, ε3, and ε4. The ε4 allele represents a crucial genetic risk factor for late-onset Alzheimer's disease (LOAD), increasing the risk for heterozygous carriers by three times and for homozygous carriers by 15 times. In early-onset Alzheimer's disease (EOAD), the risk rises in cases of people with a family history (Hoogmartens et al., 2021). Even though the ε4 allele has been associated with a higher risk, the carriers do not all develop AD, because other factors are involved in the development of the disease. Variants in the SORT1 gene are present in both familial and sporadic forms of AD, linking it to the transport of amyloid precursor protein (APP) (Nicolas et al., 2018).

3.2.1.2 Diagnosis and Treatment

When a patient already diagnosed with possible AD comes to a primary care physician, the first steps that will be taken involve checking medical and family history, reviewing medication history to verify any cognitive effects, conducting a bedside cognitive assessment (MMSE or MOCA), and performing a blood test to eliminate any reversible forms of dementia. Routine tests can include a full blood count (CBC), a full metabolic panel (CMP), thyroid-stimulating hormone (TSH), and vitamin B12 levels, which do not usually represent certain anomalies related to the presence of Alzheimer's disease. A brain CT scan can show cerebral atrophy and an enlarged third ventricle, but that is not conclusive of AD because this may also be caused by other diseases or normal ageing (Haapasalo & Hiltunen, n.d.; Petersen, 2018). Magnetic Resonance Imaging (MRI) represents a more effective method for evaluating dementia, particularly AD, by showing such features as the loss of the entorhinal and medial temporal cortices. Though MRI is not capable of a conclusive diagnosis of AD, it aids in identifying patterns of other neurodegenerative diseases (Sobański et al., 2020). Volumetric MRI measures brain volume changes, such as the shrinkage of the hippocampus of the brain due to cognitive impairment and Alzheimer's disease; however, its utility in the initial stages of detection is questionable. PET and fMRI functional imaging modalities demonstrate a promise of tracing the advancement of AD; however, their diagnostic capabilities have not been fully developed. Electroencephalography (EEG) is generally ineffective in the assessment of AD, mostly appearing normal. A neuropsychological test is needed to detect mild cognitive impairment (MCI) at an initial stage (Bottino et al., 2002). Currently, no cure has been discovered to treat AD, and the clinical practice focuses on symptomatic treatment (Adlimoghaddam et al., 2018). Two types of drugs have been approved; these include cholinesterase inhibitors and partial NMDA antagonists. The cholinesterase inhibitors, such as donepezil, rivastigmine, and galantamine, elevate acetylcholine levels; however, they may also lead to adverse effects, including vomiting, nausea, and bradycardia (Lovras et al., 2025). Memantine, a partial NMDA antagonist, has received approval for the treatment of moderate to severe AD, with reported side effects including headache and dizziness (Khoury et al., 2018). Additionally, advances in the comprehension of AD have led to the creation of disease-modifying drugs, such as Aducanemab, Lecanemab, and Donanemab, which aim at removing amyloid in the brain (Hsu & Siwiec, 2022). Aducanemab and Lecanemab were both fast-tracked by the FDA, showing results on amyloid-beta plaques and showing a difference in the degree of the disease progression suppression (Rashad et al., 2022; Kumar et al., n.d.).

3.2.2 Bell's Palsy

Bell's palsy represents the most prevalent cause of lower motor neuron facial palsy, characterized by a rapid, unilateral weakness of the face and with such characteristics as postauricular pain, dysgeusia, and sensibility of the face (Singh et al., 2022). The issue is linked to the anatomical structure of the facial nerve, where there are motor, sensory, and parasympathetic fibres, and thus they may be interconnected with other cranial nerves and additional symptoms (Alqahtani et al., 2024). The peak degree of disability manifests within the initial 48–72 h, with recovery and overall quality of life being contingent upon the severity of the condition. The exact mechanism of pathophysiology remains controversial, and some variables, including herpes simplex infection, nerve compression, and autoimmunity, have been proposed, but the mechanisms of action are not fully understood (Menchetti et al., 2021). Cranial nerve VII (CNVII) is the facial nerve, which has motor and sensory functions as well as parasympathetic functions. It makes it possible to make voluntary motions of the face, to feel taste in the anterior two-thirds of the tongue, and to control the secretions of the salivary and lacrimal glands. The nerve receives axons from the solitary nucleus and the superior salivary nucleus, providing sensory and parasympathetic input. Additionally, it incorporates motor fibres from the facial nucleus, which predominantly receives input from the contralateral motor cortex for various facial movements. This circuit is composed of intracranial, intratemporal, and extratemporal components that traverse the pathway of the pontomedullary angle, extending to the internal acoustic meatus and along the geniculate ganglion. It develops as it grows, producing significant branches, including the superior petrosal nerve and chorda tympani. The final branches, temporal, marginal mandibular, buccal, zygomatic, and cervical, are important in facial expression and operations, including the closure of the mouth and eyes (Lee et al., 2023).

Even though several immunological, infectious, and ischemic pathways have been proposed to explain Bell's palsy, its aetiology is dubious. A potential aetiology has been proposed, involving the reactivation of the herpes simplex virus (HSV-1) at the geniculate ganglion. This suggestion is supported by findings of HSV-1 in the facial nerve fluid during decompression procedures, as well as the virus's capacity to induce facial palsy in animal models (Neckel et al., 2025). Unlike other diseases associated with HSV, the palsy manifested by Bell shows distinct features and makes the cause-and-effect relationship between it and HSV-1 unclear (Wang et al., 2025). These neurotropic human herpesviruses, such as HSV-2 and varicella zoster virus (VZV), can develop latency in ganglia, remaining in that state throughout the life of the host, with no active replication. There is also the possibility of reactivation of HSV and VZV, particularly in immunocompromised individuals, but it is not always accompanied by the presence of antibodies (Canova et al., 2024). One potential pathogenesis of neurological deficit in the HSV-1-related palsy that Bell had is related to axonal destruction and apoptotic processes induced by the virus. Recent investigation highlights the importance of intra-axonal signalling molecules, mitochondrial permeabilization, and Wallerian degeneration as a natural immune response to restrict viral transportation to the central nervous system (Hill & Loreto, 2021). Research conducted in vitro has demonstrated that the peripheral nerve axons produce local messenger RNA in response to HSV-1 infection, meaning that the axon is autonomous to the virus (Hakiminia et al., 2022). It has been shown that acute changes in sodium conductivity inhibit calcium homeostasis, which leads to the activation of proteases and axonal injury, perhaps explaining the onset of Bell's palsy and the associated swelling of nerves (Esteves et al., 2022). Furthermore, a cell-mediated immune response targeting myelin, akin to that observed in Guillain-Barré syndrome, may also play a role in the disease's pathophysiology, as evidenced by alterations in blood lymphocyte proportions and myelin protein reactivity in affected individuals (Luo & Ding, 2022). Bell's palsy is diagnosed by clinical examination and is characterized by acute lower motor neuron unilateral facial paralysis that reaches a maximum severity in 72 h. The symptoms often include neck, mastoid, and ear pain; dysgeusia; hyperacusis; and loss of face sensibility and are present in 50–60% of cases, which supports the diagnosis (Marques et al., 2022). Dysfunction will have an impact on the facial nerve, which is in the temporal bone, and an MRI can show an amplification of specific portions. However, VZV and HSV PCR assays of various collections have not shown any dependable relationship with the clinical presentation, therefore limiting their diagnostic effectiveness (de Castro et al., 2022).

3.2.2.1 Various Treatment Strategies for Bell's Palsy

The most recent recommendations given by the American Academy of Neurology and the American Academy of Otolaryngology indicate the use of corticosteroids to treat Bell's palsy and discourage the application of antiviral therapy as a routine practice (Sialakis et al., 2025). The oral steroid therapy should begin within 72 h of symptom onset, as per the European randomized controlled trials (Babl et al., 2022). The lack of significant benefit of the combination of corticosteroids and antivirals can be due to the fact that small cases, which tend to disappear on their own, are included, whereas non-blinded studies support the combination therapy. The hypotheses of HSV-1 and the occurrence of misdiagnosis of zoster sine herpete (Sialakis et al., 2025) support the use of antivirals, especially acyclovir. The varicella-zoster virus (VZV) is pivotal in diagnosing acute lower motor neuron facial palsies, as Ramsay-Hunt syndrome is associated with a less favourable prognosis than Bell's palsy and shows a more positive response to antiviral and corticosteroid therapies. Valacyclovir 3000 mg/day is typically administered for 7 days in the treatment of Ramsay Hunt syndrome because it has a greater bioavailability than acyclovir (Boukhvalova et al., 2022). Acyclovir plus corticosteroids against Bell's palsy is debatable; however, corticosteroids alone are shown to work. Additional research is essential to elucidate the benefits of combination therapy in severe cases. Acyclovir may need intravenous administration in patients with severe immunocompromised conditions in order to prevent complications related to the central nervous system (Zhu et al., 2025). The following are various other treatment strategies of Bell's palsy.

- **Eye care:** Implementing an eye protection strategy is essential for patients with impaired eyelid closures to prevent epiphora, keratitis, and ulceration. The preventative measures that clinicians should use include application of wrapped sunglasses, artificial tears, ointments, and taped eyelid closure at night. Patients are not supposed to expose themselves to harsh conditions like showers, swimming pools, or areas with wind or dust. The eyelid weights could prove beneficial to the senior citizens, diabetic patients, those who have underlying eye problems, and those who have chronic pain in spite of other treatment methods (Tawfik & Dutton, 2023).

- **Oral care:** Damage to the sphincter in the orbicularis oris results in oral incontinence and heightens the likelihood of abrasions on the lips and inner cheeks during the act of chewing. Straws and soft foods are recommended in cases of flaccid paralysis of the face. The loss of the ability to evert the lower lip restricts the dietary options, but interim denture spacers on molars can guard the buccal mucosa against damage.
- **Physiotherapy:** Heat therapy, mime therapy, massage, electrostimulation, and biofeedback serve as evidence-based interventions for addressing Bell's palsy; however, an analytical challenge arises from the variability in these interventions (Ly et al., 2022). Physiotherapy benefits may be received by special subgroups, such as those with incomplete recovery or synkinesis (Fu et al., 2024). Before chemodenervation, neuromuscular retraining can be used with such patients to improve physiotherapy outcomes when dealing with synkinesis.
- **Synkinesis:** Synkinesis is a form of unusual contractions of the face muscles during voluntary action, which is, in most cases, due to the reinnervation of the facial nerves after they have been damaged. This can lead to involuntary eye closure during eating or smiling, the movement of the lips during the eye closure, or dimpling of the chin. The treatment is largely by means of physiotherapy using biofeedback exercises on their symmetry and specific botulinum toxin injections that will be tailored to each individual (Neville et al., 2022). Most of them report that they are happy and improving, and measures are taken to avoid injecting certain muscles to preserve the ability to smile. Certain people may encounter diminished efficacy from successive treatments and may necessitate injections many times annually. Surgical interventions such as selective myectomy or neurectomy may provide enduring treatments (Shikara et al., 2023).
- **Facial reanimation:** Methods to rehabilitate midface mobility, particularly for individuals with partial recovery, encompass nerve-to-nerve transfers and tissue transfers. The localized tendon transfer method created by Labbe relocates the muscle tendon of the temporalis to enhance the movement of the oral commissure (Saraniti & Verro, 2024). By contrast, the method of innervated free muscle transfer uses the gracilis muscle together with neurons provided by either the contralateral nerve or the ipsilateral nerve. Patients who show a frozen oral commissure and who still have functioning facades indicate that they prefer nerve-to-nerve transfer, and they use a donor nerve, like the masseteric branch. The processes are aimed at increasing the quality of life and maximizing the level of smiles (Eviston et al., 2015).

3.2.3 *Cerebral Palsy*

According to the definition provided by the Centers for Disease Control and Prevention, cerebral palsy is a set of conditions that causes mobility, posture, and balance impairment and is devastating due to damage in the immature brain. They are long-term clinical manifestations that are not progressive but may evolve

(Ikeudenta & Rutkofsky, 2020). The causes of cerebral palsy are primarily prenatal, with approximately 92% of cases attributed to these factors (Morgan et al., 2018). Intriguingly, some of the risk factors are preterm birth, perinatal infections (including chorioamnionitis), intrauterine growth restriction, and multiple gestation that cause brain injuries (Shi et al., 2017). Less than one out of ten develops as a result of intrapartum hypoxia, yet about 8% develop in adulthood because of head injuries or infections (Wang et al., 2025). Importantly, 80% of the cases are idiopathic, lacking an aetiology. Cerebral palsy is marked by an eclectic set of mobility problems along with the presence of such complications as poor balance and sensory problems. Common comorbidities include pain, intellectual disability, ambulation impairment, hip displacement, speech impairment, epilepsy, incontinence, and behavioural or sleep disorders. These outcomes present themselves even at a later expected stage of development and might even extend to hearing impairment, vision loss, and the appearance of scoliosis because of spasms in the muscles (Tedeschi, 2024).

3.2.3.1 Diagnosis and Treatment

Cerebral palsy (CP) is linked to specific movement deficiencies, such as spasticity, dyskinesia, ataxia, or a complicated gait. Spasticity, which affects approximately 80% of juvenile cases, can affect paediatric patients as diplegia, hemiplegia, or quadriplegia (Mendoza-Sengco et al., 2023). Modern methods of imaging, like prenatal ultrasound and MRI make it possible to diagnose it earlier, maybe as early as 6 months. According to the American Academy of Neurology, there is a diagnostic approach, which includes a patient history, physical examination, and neuroimaging, where MRI is preferred due to its high specificity in detecting cerebral abnormalities (Aravamuthan et al., 2021). Hypoxic-ischemic damage has a few specific symptoms despite the many abnormalities that can appear. If neuroimaging lacks diagnostic information, further testing for metabolic or genetic anomalies can be considered, although the evidence for this testing is limited. The severity of cerebral palsy and the effect of treatments are evaluated with the help of a variety of scales; the Gross Motor Function Classification System (GMFCS) (O'shea, 2008) is one of the most commonly used evidence-based tools. The GMFCS divides gross motor functioning into five levels, with Level I being the least and Level V being the most severe limitation. GMFCS levels of the patients may be longitudinally followed to determine the changes in the patients after treatment (Rana et al., 2025). Additional tools, including the Wong-Baker FACES Pain Rating Scale, can be used to determine the effectiveness of the reaction to treatment. Patient management in cerebral palsy depends on the type of symptoms exhibited by the patient and thus requires pragmatic goals to be set alongside those of their families. Personalized care necessitates the collaboration of a diverse team of professionals. At the age of five, kids have progressed to about 90% of their motor development, despite severe treatment (Novak, 2014). The treatment should also ensure that communication, academic, social, and professional skills are built as the kid develops. It is important to address common issues, but most interventions have a small amount of short-term data due to the challenge of studying this vulnerable group (Joint Committee on Infant Hearing, American Academy of Audiology, American Academy of Pediatrics, American Speech-Language-Hearing Association, Directors of Speech and Hearing Programs in State Health and Welfare Agencies, 2000). Following are the various treatment strategies for cerebral palsy.

- **Spasticity:** The management of spasticity in individuals with cerebral palsy focuses on the prevention of abnormalities, alleviation of pain, and preservation of functional abilities. Referrals to surgical specialists depend on the extent of the disease, with the referrals of GMFCS level V patients (referrals between ages 1 and 4) and level I patients (referrals between ages 5 and 10). The most common one is onabotulinumtoxin A (Botox), whereas data on the timing is inconclusive, and injections typically start after 18–24 months (Esquenazi et al., 2023). Systemic medications, including baclofen and diazepam, can help with a lot of side effects, limiting their use over time. Surgery can be used to improve motor control in patients classified as GMFCS level II and III; selective dorsal rhizotomy should be evaluated in 4–5 years (Pavone et al., 2016). It shows temporary improvements to gait but has inconsistent long-term functional improvements. The intrathecal baclofen is recommended in non-ambulatory children (GMFCS level IV or V) with a short-term improved quality of life despite an increase in complications and costs (Ailon et al., 2015).
- **Hip Disorders:** Children with cerebral palsy have hip abnormalities affecting about 36% of this population age group, and the occurrence escalates alongside elevated levels of GMFCS. Spasticity can cause ipsis and dislocation of the hips and result in non-ambulatory children being difficult to manage (Huser et al., 2018). Regularly, including in terms of frequency, surveillance (assessment and radiographic imaging) plays a vital role in detecting problems at the earliest stage and depends on the level of GMFCS. In the United States, the hip monitoring program is not well organized, but in Europe, Australia, and Canada, recommendations are used. A spasticity surveillance system, including prophylactic surgery, has shown a decrease in issues related to the hips.
- **Improving Movement:** Physical and occupational therapies are essential to reducing the adverse effects of mobility and balance in patients with cerebral palsy and involve various strategies, such as stretching, electrical stimulation, and endurance training. These therapies increase gait and motor function, though sufficient evidence is lacking to support one modality and the optimum treatment parameters (Branjerdporn et al., 2018). Referral to therapy should be timely when it is known, and a combination of therapy and onabotulinumtoxin A injections could improve motor functions. Online family resources and home rehabilitation programs can be used to improve patient functionality and skills, especially in the prevailing upper limb (Novak et al., 2009).
- **Improving Hand Function:** Movement treatment that is triggered by a constraint and intense bimanual therapy is effective in improving the functions of the hands of children with hemiplegia. The former focuses on using the nondominant

hand by restricting the dominant hand, and the latter proposes the involvement of both hands. The two interventions produce positive functional outcomes that could last up to 6 months after treatment (Gordon et al., 2011). Hand-arm intensive bimanual therapy could be another alternative approach for children frustrated by constraint-induced movement therapy.
- **Improving Motion:** Serial casting has been used with cerebral palsy patients to increase range of motion, although there is little evidence that these small short-term improvements result in functional improvements. Thus, this drug must be provided in cases when other medications have already been ineffective (Tustin & Patel, 2017; Vitrikas et al., 2020).

3.2.4 Seizures

Epilepsy is a significant issue of both childhood and adult neurobiological disorders; compared to epilepsy, seizures are a common disorder. Epilepsy is a disorder that comes with repeated seizures accompanying cerebral abnormalities (Shorvon et al., 2016; Nordberg et al., 2023). A seizure refers to sudden and uncontrolled changes in perception or behaviour because of the uncharacteristic release of cortical neurons. It is essential to differentiate between a seizure and the postictal period, which follows the seizure and is marked by the gradual return of normal neurological functions. This phase is typically characterized by transient impairments in cognitive, conscious, and sensory abilities. The interictal period is the phase where the person restores normal neurological functioning. Seizures may occur in many different forms, including both simple and complex forms, differing in length and change of awareness. Furthermore, they may exhibit a range of motor, psychic, affective, sensory, or autonomic signs and symptoms.

3.2.4.1 Partial Seizures

Partial seizures involve only a certain area of the cortex, but the generalized seizures affect both hemispheres (Beniczky et al., 2022). There are three types, namely simple partial seizures, where one remains conscious; complicated partial seizures, where the awareness is impaired; and generalized partial seizures, where there is impaired consciousness and tonic-clonic activity. Additionally, simple partial seizures, or auras, are localized to one cerebral lobe or hemisphere and thus exhibit specific symptoms relevant to the part of the brain (Wang et al., 2021). Seizures in the occipital lobe induce visual hallucinations, parietal lobe seizures elicit regional paraesthesia, frontal lobe seizures can provoke tonic or clonic movements, and temporal lobe seizures manifest as psychic or emotional disturbances. The seizures do not impair consciousness and are classified into four primary categories: motor signs, autonomic symptoms, sensory symptoms, and psychic symptoms. They are typically persistent for the sufferers in most cases.

- **Motor Signs:** The seizures often begin with clonic or tonic contractions in specified body parts. As they spread through the precentral gyrus, they display a procession termed a "jacksonian march" (Zawar et al., 2024). The frontal cortex is usually associated with discharges that lead to complex behaviours like head or body turns and rotational positions of the tonic limbs. The involvement of the additional motor area can trigger an adverse seizure, which is characterized by the movement of the head and eyes along with bilateral movements of the proximal limbs (Cox et al., 2022). Additional indicators of simple partial seizures may encompass speech cessation, vocal expressions, and eyelid myoclonus, originating from the frontal or occipital lobes.
- **Sensory Seizures:** Simple partial seizures that begin at the sensory cortex cause symptoms that are related to the affected lobe: parietal lobe-related paraesthesias, occipital lobe-related visual abnormalities, and temporal or frontal cortices-related aversive olfactory or gustatory sensations. Occipital or posterior temporal seizures are associated with visual hallucinations, but auditory symptoms, like hyperacusis or hypoacusis, with temporal lobe seizures (Di Giacomo et al., 2023).
- **Autonomic Seizures:** A seizure may arise from the partial engagement of limbic structures located within the mesial temporal and frontal lobes, which influence the hypothalamus and the brainstem. The common side effects include epigastric discomfort, nausea, lightheadedness, pallor, flushing, diaphoresis, piloerection, mydriasis, tachycardia, and incontinence.
- **Psychic Seizures:** Psychic simple partial seizures can develop out of the limbic and association cortex, producing similar symptoms to psychiatric illnesses, such as deja vu, jamais vu, compulsive thoughts, cognitive disorders (i.e. depersonalization, dreamlike states, and time/place distortions), and affective syndromes (i.e. fear and anger). The diagnosis typically relies on a history of characteristic transient symptoms, and at times, the presence of an abnormal interictal EEG. However, the ictal EEG also shows abnormalities only in 25 cases, which makes it unreliable as a diagnostic tool in the case of negative cases (Jaafar et al., 2022; Cao et al., 2022). Diagnosing simple partial seizures can prove challenging, particularly when they represent the only manifestation of epilepsy and the EEG results appear normal, necessitating the exploration of alternative diagnostic methods.

3.2.4.2 Complex Partial Seizures

Psychomotor seizures, or complex partial seizures, are the most common and usually start in the temporal lobe. Lasting between 30 s and many minutes, they often start with an aura and lead to a lack of consciousness (Ghulaxe et al., 2023). Involuntary movements (automatisms) are common and include such actions as chewing or lip-smacking. These movements are not only complex partial seizures but are also found in other forms of seizures (Ghulaxe et al., 2023; McGonigal, 2022). Cases of amnesia are common in these events, followed by postictal disorientation and fatigue in the patient. The changes in the electroencephalogram are usually observed in such seizures, with the exception of rare cases with origins in the deep frontal lobe, and may be accompanied by autonomic changes including tachycardia and high prolactin levels (Wang et al., 2021).

3.2.4.3 Secondarily Generalized Tonic-Clonic Seizures

Generalized tonic-clonic seizures, historically referred to as grand mal, exhibit a consistent presentation and may potentially arise from partial seizures. A tonic stage is characterized by crying and muscular spasms, followed by the clonic stage that is characterized by the relaxation of the muscle and frequent contractions of the muscles (Vieluf et al., 2021). Significant changes in physiological processes include increased heart rate and blood pressure and respiratory secretions; such phenomena as apnea and cyanosis may also appear. The tonic-clonic phase lasts an average of 1 min, followed by a prolonged period of deep sleep that is longer than the fatigue period experienced after complex partial seizures (Cerulli Irelli et al., 2024).

3.2.4.4 Generalized Seizures

The generalized seizures entail a general dysfunction of the brain activities that take place on both sides and are accompanied by loss of consciousness. Motor signs are primarily synchronized, and the EEG alterations indicate colossal neuronal release. The next section discusses six different types of generalized seizures, which include tonic-clonic, absence, myoclonic, clonic, tonic, and atonic seizures (Steriade et al., 2022).

- **Tonic-Clonic Seizures:** They are similar to secondarily generalized seizures, but such seizures occur without a feeling or a complex partial seizure. Myoclonus or absence seizures are rare in these people. There can be a generic prodrome of the seizure, which may appear a few hours prior to it, with a headache, insomnia, mood changes, or irritability. Consciousness loss is inevitable, and it is followed by a postictal period, which entails disorientation and tiredness (Rider et al., 2024). Concomitant autonomic changes that include tachycardia and high levels of prolactin are followed, and they manifest diversely, like tongue biting, urine incontinence, and postictal paralysis.
- **Absence Seizures:** Absence seizures, formerly referred to as petit mal seizures, are categorized into typical and atypical forms (Specchio et al., 2022). Idiopathic generalized epilepsies are characterized by the often-occurring incidents of unconsciousness in brief periods (less than 10 s) and are usually accompanied by unobtrusive manifestations such as clonic movements, changes in the tone of posture, and automatisms (Eadie, 2024). Common autonomic symptoms include pallor and tachycardia, and seizures are usually followed by a quick regaining of

normal cognition. They can be occasioned by hyperventilation. In contrast, atypical absence seizures are longer lasting (10–25 s), associated with motor expression and postictal disorientation, and usually occur at the time of awakening (Kerr et al., 2021). The EEG presentation varies significantly as typical seizures are characterized by symmetrical spikes and slow-wave complexes occurring at a frequency of 3–4 Hz. In contrast, atypical seizures exhibit erratic bursts of spikes and waves, typically at a lower frequency (Harvey & Shahwan, 2023).

- **Myoclonic Seizures:** Myoclonus refers to sudden, involuntary movements of the muscle, which might be of epileptic or non-epileptic causes. Myoclonic seizures are characterized by unilateral or bilateral synchronous twitches that can either occur in isolation or in rapid succession and affect some body part or the whole body (Ascoli et al., 2021). These seizures can be spontaneous or sensory-stimulus-induced, in which case they are very short and do not affect awareness. Myoclonic seizures are linked to both primary and secondary generalized epilepsies, especially adolescent myoclonic epilepsy, and generally do not exhibit a postictal state. The ictal and interictal EEGs reveal short episodes characterized by generalized spike-and-wave or polyspike-and-wave activity occurring at a frequency of 3 to 5 Hz (Eroğlu et al., 2025).
- **Clonic Seizures:** Clonic seizures typically present before the age of three and resemble tonic-clonic seizures; however, they do not encompass the tonic phase. Asymmetry may be observed, accompanied by a loss of consciousness, followed by a state of postictal disorientation (Cerulli Irelli et al., 2024). Ictal EEG patterns demonstrate pronounced high-amplitude polyspike-and-wave discharges, followed by a phase of low-amplitude slowing or generalized EEG suppression.
- **Tonic Seizures:** Tonic seizures typically begin during early childhood or adolescence and are characterized by tonic spasms that affect the truncal and facial muscles, along with flexion or extension of the limbs and changes in consciousness, lasting from 5 to 20 s (Cerulli Irelli et al., 2024). The electroencephalogram reveals a pattern of widespread low-voltage rapid activity, followed by a semi-rhythmic delta wave activity.
- **Atonic Seizures:** Atonic seizures, commonly referred to as drop attacks, are characterized by a sudden loss of muscle tone in postural muscles, often beginning in childhood. Minor instances may cause transient head dips, however severe instances might result in abrupt collapse, posing a danger of lacerations and fractures (Sirven, 2002). These events last less than 5 s and are linked to changes in awareness however, they do not always result in significant postictal effects. They are mostly observed in symptomatic generalized epilepsies, where the ictal EEG contains reduced electrical activity and is often followed by the spike-and-wave or slow-wave discharges (Mazumder et al., 2022).

3.2.5 Multiple Sclerosis

Multiple sclerosis (MS) is a long-term disease of the central nervous system that manifests itself in the immune-mediated demyelination. It is characterized by the fluctuations in time and space in inflammation and demyelination. The symptoms, which are mostly motor impairments such as stiffness and weakness, greatly influence the balance and mobility, and, as such, multiple sclerosis is one of the key determinants in the predicament being experienced by young adults. Jean-Martin Charcot was also the first clinician to note the importance of multiple sclerosis on gait, and therefore there has been a lot of research into the effects of the disease on motor impairments, which has increased in the past few years (Thompson et al., 2018). Cotsapas et al. (2018) carried out research in relation to genetic factors that cause multiple sclerosis (MS) and reported that it had a population of two per 100,000 in Japan and more than 100 per 100,000 in Northern Europe and North America (Cotsapas & Mitrovic, 2018). The incidence of multiple sclerosis is twice as high in the temperate areas in comparison to the tropical ones, women are twice as prone to it, and the highest incidence is found in whites. The disease usually reaches its peak in the fourth decade of life, but it can occur between ages 15 and 60, although it is rare to see it in childhood. The third most common cause of disability in people between 15 and 50 years of age in the U.S. is multiple sclerosis, followed by trauma and musculoskeletal diseases.

Multiple sclerosis (MS) is thought to arise from a complex interplay of various factors, including genetic susceptibility, unidentified environmental influences, and the interaction of the host immune system with the central nervous system. Genetic predisposition is significant, and the lifetime risks of siblings of the affected individuals are 2–5%, and the parents and offspring 1%. The monozygotic twins have a risk rate of 30%, and the dizygotic twins have a similar risk like siblings. The early genetic research focused on specific genes that are related to the probability of multiple sclerosis, such as the human leukocyte antigen (HLA) and other immunological variables. However, such studies often yielded conflicting results due to limited sample sizes and poor understanding of the pathogenesis of diseases since many of the genes of interest have no clear variations in disease risks (Terwilliger, 2022). Genetic linkage studies utilizing microsatellite markers have revealed that the major histocompatibility complex (MHC) and various immunoregulatory genes significantly contribute to hereditary risk (Malviya & Rajput, 2025). The follow-up genome-wide linkage analysis conducted on consanguineous families validated the presence of HLA risk alleles; however, it revealed no significant linkage to regions beyond the MHC (Salehi et al., 2024). The International Multiple Sclerosis Genetics Consortium (IMSGC) conducted a comprehensive multinational study involving 4506 single-nucleotide polymorphisms across 730 multiplex families. The study established significant linkage within the HLA region and identified several suggestive linkage peaks throughout the remaining portion of the genome (Goris et al., 2022).

Genome-wide association studies (GWAS) have been effective in determining risk alleles of multiple sclerosis (MS) through an examination of the variations in allele frequencies in cases and controls. There was also a significant finding of the association of certain single-nucleotide polymorphisms (SNPs), in particular, IL2RA and IL7RA genes that have been implicated in type 1 diabetes and Graves' disease and thus confirms that multiple sclerosis is a genetic autoimmune disease (Graves et al., 2023). However, barriers continue to exist in the functional explanation of GWAS results and genetic mapping of secondary clinical characteristics because of limitations in data. The common variant hypothesis assumes that prevalent genetic variations may be limited to a small number of common variants that are believed to contribute to illness risk and account for a small percentage of the total population risk (Shirai & Yamauchi, 2022). Based on migration studies, it is shown that profound environmental exposure that takes place in relation to multiple sclerosis (MS) occurs well before the age of 15, and that patients who live with increased distances to the equator exhibit reduced vitamin D concentrations and increased susceptibility to MS. (Läderach & Münz, 2022) The relationship between multiple sclerosis (MS) and latent Epstein-Barr virus (EBV) infection remains a subject of debate. Some studies indicate that EBV may play a role in immunological activation within MS lesions, particularly among large cohorts of young adults. Conversely, the presence of EBV is not consistently detected in the brains or cerebrospinal fluid of individuals with MS. Biological sex is the most dominant risk, whereby over three-fourths of the patients are female, and the reasons for this is the case are not well understood (Bjornevik et al., 2022; Soldan & Lieberman, 2023). Most multiple sclerosis follows a relapsing-remitting course, and approximately half of them progress to secondary progressive multiple sclerosis, and only 5% have a primary progressive form. However, the factors that influence such progression patterns and the inconsistency in the level of symptoms among patients are not well studied.

3.2.6 Clinical Motor Dysfunction

Motor dysfunction in multiple sclerosis (MS) occurs in the form of stiffness, weakness, tremor, and ataxia. Spasticity is an increase in muscle velocity-related muscle tone due to the destruction of inhibitory impulses between the corticospinal tract and gamma motor neurons and interneuron connections. Weakness, which affects up to 89% of multiple sclerosis, can be localized and linked to upper neuron motor disease when it occurs as a result of impairment of the input to alpha neurons in place of the input to lower neurons. Lesions in the cerebellum and related pathways in the brainstem and the basal ganglia, including the Mollaret triangle of cerebellar structures, are associated with tremor and ataxia (Marrie et al., 2023). In some instances, proprioceptive loss can be a major cause of disability. The brain circuits associated with tremor etiology are intricate, given the diverse outcomes of interventions like thalamotomy. Fatigue, characterized by a diminished ability to generate force during extended periods of motor activity, results in motor dysfunction in individuals that do not exhibit other motor impairments (Ghosh et al., 2022). Spinal cord and brainstem lesions or multiple sclerosis may cause paroxysmal tonic seizures, which are the involuntary contractions of limbs. These cramps, which can be caused by movement or hyperventilation, can occur on one side of the body and last from 30 s to a minute, but they are episodic, and the cramps are preceded by sensory complaints on the opposite side. Importantly, EEG at such events usually shows normal outcomes (Wood et al., 2025).

In MS, the manifestations of the condition differ based on the specific locations of demyelinating lesions within the central nervous system. Many manifestations encompass difficulties in vision, such as lack of afferent pupillary response and diplopia because of optic neuritis, and brainstem-related injuries, such as internuclear ophthalmoplegia, facial paralysis, and vertigo. Lesions of the cerebellum cause ataxia, whereas lesions of the autonomic system cause bowel and bladder dysfunctions. Patients can experience sensory impairments, such as paraesthesia. White-matter lesions are associated with cognitive impairments, and seizures are seen in about 5 per cent of cases, often associated with active lesions in the cerebral cortex. The most common upper motor neuron defect is spastic paraparesis, the consequences of which are rigidity, cramping, and weakness. The problem of gait and postural instability is imminent because many patients with MS have reduced walking speed and ability. The prognostic factors suggest that sensory symptoms will follow a benign course of disease, whereas motor symptoms, early relapses, and an age at onset over 40 years suggest a more aggressive disease progression (Gouider et al., 2024; Wergeland et al., 2023).

3.2.6.1 Diagnosis and Treatment

The diagnosis of Multiple Sclerosis (MS) is fundamentally reliant on the clinical manifestation of lesions within the central nervous system (CNS), alongside MRI findings that confirm the recurrence of these lesions in designated areas. Multiple sclerosis is typically a relapsing-remitting disease and is marked by relapses that bring in new symptoms, which appear within 24 h and can continue to persist 1–2 months before remission. These relapses are interchanged with periods of stability known as remissions. The rate of relapse thus reduces over time; however, some 10–15 years later some patients experience a shift towards a progressive course where they experience symptoms of progressive worsening of their neurological functions, such as gait and cognitive issues, known as Secondary Progressive MS (SPMS) (Cameron & Nilsagard, 2018). On the other hand, some people have the Primary Progressive MS (PPMS) type at the very beginning of the disease with no early relapses. The negative effects of MS, including motor dysfunction and weakness, are a consequence of impaired nerve conduction due to the demyelinating process and axonal damage, which lead to the gradual worsening of functional deficiency. In multiple sclerosis (MS), myelination is demyelinated, which causes positive symptoms, such as painful spasms and aberrant discharges of nerves, and negative ones that cause a gradual decline in motor performance. The MRI plays an essential role in the diagnosis of MS, the lesions of which are mainly located in periventricular and juxtacortical white matter and are usually characterized by an oval or elliptical shape (Noteboom et al., 2024). Traditional MRI methodologies, such as dual-echo and fluid-attenuated inversion recovery (FLAIR), demonstrate considerable sensitivity in identifying lesions associated with multiple sclerosis; however, they fall short in providing specificity regarding the diverse clinical substrates involved. The images obtained using gadolinium are able to differentiate between active and inactive lesions, and a darker image over a particular lesion indicates more damage. The correlation between MRI and clinical manifestations in multiple sclerosis is weak due to the problems of specificity and the inability to assess the damage or the change of the whole central nervous system functions after the injury (Beyrle et al., 2025).

It is also critical to evaluate atrophy in Secondary Progressive Multiple Sclerosis (SPMS), and Relapsing-Remitting Multiple Sclerosis (RRMS), as it is the evidence of the axonal degeneration, which may lead to irreversible damage. Various methods can be used to measure tissue volumes; most of them are reproducible and sensitive to change, but the accuracy varies across different methods (lacking a gold standard) (Gajewski et al., 2025). One of them showed that there was continuous brain atrophy in multiple sclerosis patients, regardless of the subtype. The McDonald criteria have been found to be vital in diagnosing multiple sclerosis, where MRI data is vital in enhancing specificity and sensitivity, particularly in early diagnosis and response to treatment (Montalban et al., 2025). Revisiting these criteria has refined the diagnostic requirements for Primary Progressive MS (PPMS) to include specific MRI abnormalities alongside evidence of disease progression (Fissolo et al., 2025). The International Panel on Diagnosis of Multiple Sclerosis has revised the 2010 McDonald criteria, contributing to the establishment of the 2017 diagnostic criteria for Multiple Sclerosis (MS). The revisions improve the diagnosis of patients with clinically isolated syndrome (CIS) by focusing on the detection of lesions of the central nervous system that do not have other symptoms to explain. The major changes involve replacing oligoclonal bands (OCBs) in cerebrospinal fluid with dispersed in time (DIT) and assessing both symptomatic and asymptomatic lesions identified through magnetic resonance imaging. Multiple sclerosis is now diagnosed following a solitary clinical incidence under specific lesion requirements. These criteria need to be refined by further investigation, especially in heterogeneous populations, to prevent misdiagnosis, and a method of differential diagnosis is needed to avoid misdiagnosis that occurs due to different imaging modalities (Rovira & Auger, 2021).

Magnetic resonance spectroscopy (MRS) is used to detect signals in organic molecules rather than protons in water, unlike magnetic resonance imaging (MRI). This technique helps to measure neurodegeneration and neuronal integrity through such indicators as N-acetylaspartate (NAA) and such metabolites as choline and GABA. Lymphocytic pleocytosis in the cerebrospinal fluid (CSF) and the detection of oligoclonal bands (OCB) are frequently observed in individuals diagnosed with multiple sclerosis (MS). Nonetheless, the findings do not always indicate an active disease process but may as well be evident in numerous other neuroinflammatory conditions (Adamson et al., 2024). The non-invasive methods, including transcranial magnetic stimulation (TMS) and functional MRI (fMRI), study the synaptic transmission and the reaction of the brain to the damage. The transcranial magnetic stimulation (TMS) of the primary motor cerebral region can elicit an electric current in the neurons, which allows testing of the functionality of the motor mechanism and the diagnosis of normal and pathological conditions (Antczak et al., 2021). TMS can be used to observe the plasticity of the brain, and it implies that the centre of excitability may be moved to a new location that may not align with the regular boundaries after brain damage such as demyelination. It is possible that either the stimulation of dormant hotspots or the stimulation of dormant synapses or the establishment of new synaptic connections is causing the movement (Zhang et al., 2021). It is possible that unilateral lesions may cause changes in the interhemispheric output maps, and this is the situation in multiple sclerosis (MS), where the motor cortical area of the affected hemisphere tends to swell, and that is coupled with an improvement in the condition. Observations of neural plasticity are also supported by changes in parameters of MEP and transcallosal connections after lesions (Barbi et al., 2022). The advancement of MRI technology has significantly enhanced the understanding of the dynamic alterations in brain tissue and the interconnections within networks as they respond to motor behaviours. An investigation that analysed the relationship between fMRI activity and T2 lesion volumes in non-disabled RRMS patients showed increased cortical activity in several brain areas when performing simple activities, which showed potential mitigation of the functional effects of the white matter damage. Similar patterns of plasticity were identified in SPMS patients, indicating augmented adaptive changes in specialized cortices in the course of motor task execution, which also helps to counter the effects of MS-related damage (Mistri et al., 2023).

The evoked responses are important in identifying the subclinical lesions as well as linking symptoms with the CNS mechanisms. Trimodal stimulated reactions, namely visual, auditory, and somatosensory, are usually used in medical practice to detect lesions caused by demyelinating diseases. Long P100 latency Visual evoked responses (VER) are indicative of prechiasmal optic nerve conduction block in a significant proportion of the patients with optic neuritis (Naseri et al., 2021). Brainstem auditory evoked responses (BAER) may exhibit extended latencies and absent waveforms in individuals with confirmed lesions. Somatosensory evoked responses (SSER) exhibit great sensitivity for MS, particularly in spinal cord lesions. Older multiple sclerosis patients often demonstrate gradual progression, with initial attacks infrequently occurring after age 60; alternative illnesses must be excluded in these instances. Notably, complete normal MRI findings are not characteristic of multiple sclerosis, even of recurrent cases of isolated optic neuritis (Younger, 2023).

3.2.7 *Neurofibromatosis*

Neurofibromatosis (NF) is a hereditary condition that is characterized by the occurrence of tumours on nerve tissues, which are classified in three different types: NF1, NF2, and schwannomatosis (SWN) (Evans et al., 2018). The most frequent one is neurofibromatosis type 1 (NF1), constituting 96 per cent of all cases, and is characterized by the occurrence of neurofibromas, which cause a number of cutaneous and skeletal abnormalities (Kresak & Walsh, 2016). NF2, present in 3% of cases, results in auditory impairment and vestibular complications, whereas SWN, occurring in less than 1% of cases, induces intense pain. The types arise from various mutations, and presently, no preventive or therapeutic treatments exist (Campian & Gutmann, 2017).

3.2.7.1 Neurofibromatosis Type 1

Neurofibromatosis type 1 (NF1), also known as von Recklinghausen's disease, is distinguished by a variety of light-brown skin lesions, commonly referred to as café-au-lait spots, along with freckling in skin folds, noticeable neurofibromas, and the presence of Lisch nodules (Ruggieri & Huson, 2001). It affects 1 in 3000 or 4000 per 100,000 people across the globe and is also spread through autosomal dominance, and 50% of the mutations are de novo (Stephens et al., 1992). The disease advances through childhood, and the life expectancy of the affected individuals is reduced by 15 years in comparison to the normal population and with significant associations to malignant tumours and blood vessel diseases, which cause deaths of affected people before the age of 40 years (Landry et al., 2021). The NF1 gene, situated on chromosome 17 (17q11.2), is implicated in neurofibromatosis type 1 (NF1), with point mutations constituting 90% of instances, and single exon or complete gene deletions account for 5–7%. NF1 encodes neurofibromin, a Ras-GTPase-activating protein synthesized in neuronal and Schwann cells (Scheffzek et al., 1998). A lack of neurofibromin leads to the hyperactivation of Ras, subsequently triggering the activation of the AKT/mTOR and Raf/MEK/ERK signalling pathways. This activation affects neurotransmitter release and the formation of neurofibromas, but the processes underlying café-au-lait spots and learning problems linked to NF1 mutations remain ambiguous (Luna et al., 2021). A multitude of pathogenic NF1 variations has been identified in patients, with particular variants linked to severe phenotypes such as malignant peripheral nerve sheath tumours (MPNSTs), plexiform neurofibromas, and other complications. Certain mutations, such as c.2970-2972 del AAT (p.M992del), are linked to lesser symptoms, but missense variants in particular codons are connected with severe problems and characteristics akin to Noonan syndrome (Tong et al., 2022; Napolitano et al., 2022).

NF1 is associated with certain genetic defects, and it is a negative regulator of the RAS gene products that are activated in myelodysplastic syndromes and acute myeloid leukaemia. Genetic testing is used to verify diagnoses and help in screening of family members; however, a negative outcome does not rule out the possibility of the disease entirely, as there is a chance of the presence of mosaicism. NF1 is highly variable in its clinical manifestations in individuals, which makes it difficult to predict and treat the disease. Allelic variation may elucidate this variety, and a verified NF1 mutation does not signify illness severity. Targeted testing generally detects mutations, although innovations in next-generation sequencing have enhanced NF1 detection. Prenatal testing may be performed with the help of either amniocentesis or chorionic villus sampling in case a mutation in a relative has been identified (Peduto et al., 2023). The main treatment of neurofibromas is surgery; however, in cases with a plexiform type of the disease, the recurrence rate after surgery is high. The treatment of cancers caused by NF1 is not yet agreed upon due to the complex growth pathways in this situation. Anti-Ras therapy, in particular, has shown promise, and anti-Ras drugs like tipifarnib, pirfenidone, and selumetinib are now undergoing clinical trials. Selumetinib was sanctioned by the FDA in 2020 to treat children with inoperable plexiform neurofibromas. NF1 patients treated with low-grade optic glioma are usually accompanied by the use of chemotherapy that involves the use of carboplatin and vincristine (Mukhopadhyay et al., 2021; Dal Bello et al., 2023).

3.2.7.2 Neurofibromatosis Type 2

NF2 can be described as the development of an autosomal dominant disease that predisposes individuals to multiple cancers of the nervous system, which are frequently caused by de novo mutations resulting in mosaic expression. The determination is contingent upon the diagnosis derived from clinical examination and neuroimaging findings. The incidence of the disease stands at one in 25,000, with a 15-year survival rate post-diagnosis and an average age of mortality recorded at 36 years. NF2 is associated with the development of bilateral vestibular schwannomas, as well as various tumours in the brain and spinal cord. Typically, patients present with meningiomas and ependymomas at a relatively young age (Fernández-Méndez et al., 2023; Marinelli et al., 2022). NF2 patients are very likely to have aural impairment, balance problems, dermal flaps, and muscular atrophy. It may cause mononeuropathy (particularly of the facial nerve) and polyneuropathy, which may be of considerable severity in 3–5% of adults. Cataracts, optic nerve meningiomas, and retinal hamartomas are often attributed to visual impairment. About seven out of ten exhibit cutaneous manifestations, with only 1 in 10 having over 10 skin tumours. Lesions can have pigmentation and hair, subcutaneous nodules can be seen along peripheral nerves, and intracutaneous schwannomas can also occur (Forde et al., 2021; Armentano et al., 2023). The genetic mutation that relates to neurofibromatosis type 2 (NF2) is the presence of a merlin gene mutation in chromosome 22 at 22q12.2, which controls numerous proliferative signalling pathways. Merlin exerts its influence by inhibiting integrin and tyrosine receptor kinase signalling at the membrane level, while concurrently suppressing a variety of downstream signalling pathways, including p21-activated kinase, FAK/Src, Ras/Raf/MEK/ERK,

PI3K/AKT, mTORC1, Rac/PAK/JNK, and Wnt/-catenin. Merlin modulates the E3 ubiquitin ligase CRL4DCAF1 in the nucleus, thereby influencing the expression of integrins and tyrosine receptor kinases. Also, Merlin has been noted to be a key regulator of the Hippo signalling pathway that is crucial in tissue homeostasis (Sadler et al., 2021; Chen et al., 2022).

A de novo mutation results in mosaic expression, complicating the diagnosis of the tumour's tissue when it remains unexamined. Schwannomas arise as a consequence of the inactivation of both copies of the NF2 alleles, potentially through the complete deletion of the NF2 gene along with substantial segments of chromosome 22. Instances of mutations encompass protein-truncating mutations, splice site mutations, and missense mutations, with truncating mutations being the most harmful in relation to the disease. A truncated protein correlates with an earlier age of diagnosis and an increased prevalence of meningiomas and spinal tumours. Alterations in the NH2-terminal region of merlin proteins correlate with the early onset of tumour formation. Frequent modifications include splice site and silent mutations in exons 1–8, while other mutations result in less severe clinical manifestations. Mutations located near the terminal region of the gene correlate with a milder illness phenotype. Truncating mutations correlate with the Wishart phenotype, leading to more severe outcomes, whereas missense mutations located at the gene's 3′ end indicate a more favourable prognosis typical of the Gardner phenotype (Bachir et al., 2021; Forde et al., 2021). The relevance of screening for mutation in NF2 among people with unilateral schwannoma of the vestibule remains uncertain. Routine screening for germline mutations is generally not recommended, except for individuals under the age of 30; nonetheless, options such as prenatal diagnosis and pre-implantation genetic diagnosis are available (Douwes et al., 2025). No effective treatment exists for NF2 individuals, as tumours frequently recur after surgery. Treatment is advised in cases of potential brainstem compression, auditory impairment, or facial nerve complications. Stereotactic radiosurgery demonstrates effectiveness in treating NF2 schwannomas; however, it is generally not recommended for numerous or large tumours, as it may lead to vestibular impairment and trigeminal neuralgia. Although malignant transformation following radiosurgery is infrequent, subsequent surgical reoperation after such intervention may present increased difficulties (Tosi et al., 2022; Pialat et al., 2021).

Vascular endothelial growth factor (VEGF)-A is essential for schwannoma proliferation, predominantly via the VEGF-A/VEGF receptor pathway. Bevacizumab treatment, a monoclonal antibody targeting VEGF-A, demonstrates tumour reduction and auditory enhancement in more than 50% of NF2 patients with advancing vestibular schwannomas. The majority of the people are stable in hearing, as bevacizumab is observed to be the main option in the treatment of the malignancies that grow rapidly. A meta-analysis of eight studies showed that 41% experienced partial regression, 47% no change, and 7% progression with a median treatment period of 16 months. Hearing outcomes included a 20% improvement, 69% with no change, and a 6% decline. In 17% of the people, there was severe toxicity, which included amenorrhea (70%), proteinuria (43%), and hypertension (33%). The recommended dose is 5–7.5 mg/kg every 2 to 3 weeks with a treatment of 6 months, and then maintenance medication; higher dosing (10 mg/kg) does not have notable benefits but increases the chances of renal impairment (Eminowicz et al., 2012; Morris et al., 2016). Problems associated with the administration of bevacizumab include that the drug requires frequent dosages, high blood pressure, thrombosis, and resistance to drugs. In addition, it has been noted that tumour growth reoccurs after the withdrawal of treatment. A recent clinical trial of a VEGFR1/2 peptide vaccination in patients with progressing NF2-derived schwannomas showed enhancements in auditory function and reductions in tumour volume (Ouerdani et al., 2016; Tamura et al., 2019).

The NF2 gene is involved in multiple cellular growth pathways. Everolimus, a mTORC1 inhibitor, has demonstrated efficacy in NF2 patients with progressive vestibular schwannomas. Lapatinib demonstrated efficacy in 4 of 17 patients, while erlotinib was found to be ineffective according to a retrospective analysis. Mirdametinib (PD-0325901), a potent inhibitor of MEK1 and MEK2, demonstrated a notable reduction in tumour size. Hypoxia correlates with diminished progression-free survival in NF2 schwannomas, suggesting the promise of HIF-1-targeted therapy in complex cases. Cochlear or brainstem implants present potential benefits for individuals experiencing significant hearing loss, while psycho-social therapies delivered via teleconferencing have been explored for deaf patients diagnosed with NF2 (Plotkin et al., 2010; Neff et al., 2007). Patients with NF2 exhibit an earlier onset of meningiomas, which are more frequently atypical or anaplastic in comparison to sporadic cases. The long-term follow-up is still not satisfactory even after radiation therapy has been given. Investigational targeted therapies have shown some promise; lapatinib showed intermediate activity in NF2-associated progressive meningiomas, but bevacizumab showed little activity. Recent molecular studies are triggering clinical trials that examine mTORC1 inhibitors, including everolimus, with other alternative treatments, including EGFR inhibitors, PDGF inhibitors, and histone deacetylase inhibitors on NF2 malignancies (Wentworth et al., 2009; Merker et al., 2012).

3.2.7.3 Schwannomatosis

Schwannomatosis (SWN) represents a rare variant of neurofibromatosis characterized by the presence of multiple schwannomas without the occurrence of bilateral vestibular schwannomas, affecting approximately 15–20% of cases (Carter et al., 2012). The average number of tumours per patient is four, and the total tumour volume is 39 mL. It has an incidence of 11 in 40,000 to 11 in 1.7 million and is usually diagnosed at the age of 40. Symptoms such as chronic pain, numbness, and weakness may manifest early in adulthood, independent of the tumours' size or location. The average life expectancy is typically within expected ranges; nonetheless, the likelihood of developing malignant peripheral nerve sheath tumours (MPNST) necessitates additional investigation. There are not many problems with learning, and the meningioma occurs in not more than 5% (van den Munckhof et al., 2012). Approximately 85% of all cases of schwannomatosis (SWN) in families and up to

40% of all cases occurring sporadically have been attributed to germline mutations that inactivate the tumour suppressor genes SMARCB1 and LZTR1. However, these mutations do not sufficiently cause SWN, and additional somatic mutations are necessary (Hadfield et al., 2008). SMARCB1 and LZTR1 genetic testing is available. Genetic alterations, including the deletion of a copy of chromosome 22 and NF2 gene mutations, are common in Schwannomas associated with SMARCB1 mutations and support a four-hit, three-step carcinogenesis model (Hadfield et al., 2010; Rousseau et al., 2011). This entails the preservation of the mutated SMARCB1 gene within tumours, the deletion of the wildtype SMARCB1 gene copy alongside the NF2 gene, and a subsequent mutation in the remaining NF2 gene copy. Germline SMARCB1 mutations can lead to an inherited susceptibility to atypical teratoid/rhabdoid tumours (ATRT) within the central nervous system (CNS). The differentiation of SWN from isolated schwannomas can be facilitated by the mosaic loss of SMARCB1 expression, and specific morphological characteristics may suggest an underlying condition (Eaton et al., 2011; Caltabiano et al., 2017).

The tumour suppressor gene LZTR1 may also be involved in the predisposition of schwannomatosis (SWN) in the patient without SMARCB1 mutations. LZTR1 is located on chromosome 22q11.21, near SMARCB1 and NF2. In a study examining patients without mutations, LZTR1 mutations were discovered in 38% of familial cases of SWN and in 22% of sporadic instances. Somatic mutations in LZTR1 have been identified across a range of malignancies. Biallelic mutations in SMARCB1 or LZTR1 are observed in patients with schwannomatosis; however, the traditional two-hit model is unable to fully explain the proliferation of schwannomas, particularly since NF2 is often inactivated in these tumours. SWN is probably tumorigenic due to the loss of heterozygosity in large areas of chromosome 22q, which include SMARCB1, LZTR1, and NF2, (Smith et al., 2015; Frattini et al., 2013) Existing medical therapy aimed at SWN are inadequate, with medicines such as gabapentin, pregabalin, short-acting opioids, and nonsteroidal anti-inflammatories noted for their pain-relieving effects. Adjunct therapy can include tricyclic antidepressants such as amitriptyline, serotonin-norepinephrine reuptake inhibitors like duloxetine, and antiepileptic medications including topiramate and carbamazepine. If pain management is ineffective, surgical interventions, including the removal of painful schwannomas, may be considered. Surgical intervention may be required for spinal cord compression or organ impingement. The correlation of SMARCB1 and LZTR1 with histone deacetylase 4 suggests potential avenues for therapeutic intervention, particularly as histone deacetylase inhibitors are currently under development as anticancer agents (Emanuele et al., 2011; Tamura, 2021).

3.2.8 Parkinson's Disease

The diagnosis and treatment of Parkinson's disease requires the development of handicaps in day-to-day activities. Difficulties with handwriting can be followed by preliminary consultations, whereas the necessity to use antiparkinsonian medicine is often defined by functional status. Clinicians are able to observe the progress of symptoms and the impact of the increasing impairments with great ease. The challenge of disability in Parkinson's disease (PD) is linked to the severity and quality of life, as assessed at the University of Maryland PD and Movement Disorders Center through instruments such as UPDRS (Członkowska et al., 2025), OARS, and SF-12v2 (Yokoi et al., 2025). Investigations involving approximately 800 individuals diagnosed with Parkinson's disease indicate a noteworthy correlation between disability and the severity of the disease ($r = 0.64$, $P < 0.001$), as well as with quality of life ($r = -0.46$ to -0.62, $P < 0.001$) (Haddad et al., 2021). However, there is variability in the case of the individual data, such that patients with no reported disability may have a UPDRS III score between 5 and 40, and those with a score between 30 and 40 may have a subjective OARS rating between no disability and significant loss of independence, which brings up a complex interrelationship between the two metrics.

When evaluating specific requirements during rehabilitation and treatment, it is necessary to select appropriate outcome measures. Such general assessments as UPDRS might not adequately reflect the benefits of such interventions as rehabilitation programs or new pharmacological therapies for sleep disorders. On the other hand, a daily life assessment and a symptom-focused assessment of such conditions as sleep issues are more effective. Global deficits might respond to dopaminergic medication like L-dopa, but overall disability might not be affected in cases of little baseline activity or in cases with severe symptoms, like freezing of gait. Therefore, disability and quality of life might not witness significant improvements even with the reduction of some symptoms as reflected in the scores of UPDRS. PD disability assessment, often incorporated into therapeutic trials, has been evaluated using the assessment of physical disability; however, more commonly used disability measurement instruments, such as Schwab and England Activities of Daily Living (S&E) and UPDRS Part II, are flawed (Schwab et al., 2023). In contrast to the UPDRS Part II, which emphasizes impairments while neglecting particular activities of daily living (ADLs), the S and E offers a more comprehensive assessment of dependency. IADLs, which unite activities like shopping and preparing meals, are not commonly addressed in the studies on Parkinson's disease (PD), but since they determine early disability and indicate dependency requiring social support, they would be useful in the studies.

The risk factors associated with disability in PD include a notable correlation with the overall severity of the disease, the presence of depression, motor fluctuations, dystonia, and a later age of onset (Evidente et al., 2025). According to research studies conducted at the University of Maryland PD Center, it has been established that motor symptoms, which include impairment of gait, postural instability, and bradykinesia, as well as nonmotor symptoms, which include cognitive dysfunction, psychotic ideation, urinary incontinence, motivational impairments, and depression, are all contributing factors to disability (Zoila et al., 2025). The disability has no associations with tremors, and its existence underscores the lack of similarity like tremors in the case of PD. Researchers examine the development of disability in PD by establishing the correlation between the severity of the disease and the limitations in Activities of Daily Living (ADLs) and Instrumental Activities of Daily Living (IADLs) in 658 participants (Ou et al., 2021). The Unified Parkinson's Disease Rating Scale (UPDRS) was used to determine the severity of the disease, and impairment was measured using the OARS. The results indicated that the major challenge that was particularly with ambulation was linked to initial UPDRS scores (total UPDRS <20). Gait activities (dressing and shopping) were affected at an UPDRS score of 20 to 40, and more cognitively challenging instrumental activities of daily living (IADLs), such as medication taking, were associated with the severity of the disease (total UPDRS 50 to 60) (Hall et al., 2025). The significance of gait and balance in relation to the disabilities associated with Parkinson's disease (PD) is paramount. The connection between ambulation loss and disability is evident by the grouping of data points in the scatterplot based on gait or balance impairment.

3.2.9 *Motor Neuron Disease*

Motor neuron disease (MND) is a pattern of degeneration of the motor system, which affects the neurological system at a rate of two to three cases per 100,000 population (Traxinger et al., 2013). Diagnosis of it is clinical and requires upper and lower motor neuron (UMN, LMN) symptoms. MND exhibits clinical variability characterized by four distinct phenotypes: (1) amyotrophic lateral sclerosis, which accounts for 70% of all cases and has a median survival of 3 years; (2) progressive bulbar palsy, presenting with dysphagia and dysarthria; (3) progressive muscular atrophy, marked by lower motor neuron signs exclusively; and (4) primary lateral sclerosis, defined by upper motor neuron signs only and a slower progression (Hobson et al., 2016). Accumulated evidence shows that MND involves other dysfunctional systems in addition to motor neuron degeneration, such as autonomic, cognitive, cerebellar, sensory, and basal ganglia dysfunction (Orrell & Guiloff, 2023). Genetic advances have uncovered over 25 mutant genes, especially the C9ORF72 mutation, that lead to both MND and frontotemporal dementia (FTD). The mutation occurs in 40% of familial and 10% of sporadic instances of MND, in 25% of familial and 6% of sporadic cases of FTD, as well as in a range of other neurodegenerative disorders. Moreover, the nonmotor factors in the pathogenesis of MND are worth mentioning (Majounie et al., 2012). The major cause of gait, balance, and fall-related challenges experienced in MND is the progressive deterioration of both the motor and extramotor systems, along with the difficulties arising from multifaceted activities.

3.2.9.1 Gait in Motor Neuron Disease

Alterations in gait rhythm are the first indication of a person with MND, and gait cycle abnormalities appear throughout the progression of the illness. The gait cycle, which is the measure of the posture of the individual and his/her balance, involves the necessary measurements, with the most important being stride time, which is the time taken between the initial contact of the foot with the ground and the next contact with the ground (Radovanović et al., 2014). A single gait cycle comprises two distinct phases: the stance phase, during which the foot maintains contact with the ground, accounting for roughly 60% of the cycle, and the swing phase, wherein the foot is elevated from the ground, representing the remaining 30 per cent of the cycle. In healthy people, gait cycle length is not fixed and is characterized by the magnitude of fluctuations (stride-to-stride variance) and the dynamics of fluctuations (change of stride not depending on variance). This variation can be minor, at the scale of 2%, and of a subtle structure in dynamics (Sugavaneswaran et al., 2012). Hausdorff et al. (2000) identified three significant distinctions in gait rhythm between patients with MND and healthy controls: an increased magnitude of fluctuating gait, altered dynamics of fluctuating gait leading to disorganized movement, and a rise in average stride time coupled with a decrease in walking speed (Hausdorff et al., 2000). The gait alterations became evident later in the disease progression following dynamic alterations. Additional data was that of longer swing times and periods of stance, in addition to shorter stride lengths. Gait measures varied based on the location of illness development; spinal-onset MND showed a significant difference in stride duration, while bulbar-onset MND did not exhibit the same variation. On the other hand, the decrease in the length of the stride was evenly distributed across all phenotypes (Al-Chalabi et al., 2017). These gait impairments are attributed to augmented motoneuronal failure affecting upper and lower motor neurons.

Lower motor neuron dysfunction contributing to gait problems in motor neuron disease: The anomalies associated with MND and LMN are marked by weakness, muscle atrophy, and fasciculations, leading to abnormalities in the gait cycle as a result of muscle wasting and diminished endurance (Dharmadasa et al., 2016).

- **Muscle Weakness:** The MND is significantly aggravated by the loss of gait due to the progression of muscle weakness caused by the degeneration of lower motor neurons. It has been shown that muscular weakness is strongly related to gait velocity: with little muscle weakness, patients have more or less normal gait velocities, but with greater weakness, they have slower velocities (Hausdorff et al., 1997). This association has since been confirmed in a study that emphasized the presence of a negative correlation between manual muscle testing scores and performance times in the Timed Up and Go (TUG) test, the findings of which have been found to be consistent over a 6-month period of time (Montes et al., 2007).
- **Head Drop:** Flexion of the neck and axial weakness are some of the rare symptoms of motor neuron disease (MND) that are only observed in approximately 2% of cases. Dropped-head syndrome, characterized by weakness in the extensor muscles of the neck, mostly occurs as a complication of many neuromuscular conditions. It is important to note that the incidence rate of motor neuron disease is 1.3 (Gourie-Devi et al., 2003). The syndrome affects the elevation of the chin, leading to significant impairment and an increased risk of falls because of altered

spinal position and distribution of weight. It is aggravated by the fact that the handicap associated with this problem is exacerbated by poor forward visibility (Martin et al., 2011).

- **Muscle Fatigue and Loss of Endurance:** Maximal voluntary contraction may no longer be maintained, and this condition is known as physiologic muscle exhaustion, which is often associated with the dysfunction of LMN. The syndrome is quite typical of people with MND, and it leads to fatigue and lack of stamina, which adversely affect walking pace and increase the length of the stride (Vucic et al., 2011). The discrepancies in stride time and variability can be observed at the initial stages of ambulation, even though the muscular fatigue related to the activity involving lower motor neurons (LMN) can be minimal, which means that other factors, including the impaired functioning of upper motor neurons (UMN), can also cause gait problems.

Upper motor neuron dysfunction contributing to gait problems in motor neuron disease: The implication of upper motor neuron involvement in motor neuron disease is pyramidal weakening, reduction in dexterity, and muscle rigidity, which together cause gait impairments. Concomitant lower motor neuron atrophy makes clinical identification of upper motor neuron characteristics difficult. In the case of pyramidal weakness, the weakness appears more at the extremity of the lower limbs, especially affecting the dorsiflexion of the ankle and hip flexion, and the resultant loss of fine motor control in the upper limbs. These symptoms vary in severity depending on the volumes of upper motor neuron impairment among different phenotypes (Swash, 2012).

- **Spasticity:** Hyperexcitability of the spine leads to spasticity due to increased resting tone and augmentation of tone by stretch during passive movement of the limbs. The phenomenon of the clasp-knife represents inhibition that outsmarts the stimulation caused by stretch and causes sudden relaxation in the muscles. These are the symptoms that are typical of MND and are usually associated with stiff limbs and intense cramping, which affect locomotion and movement. These are significant aspects of primary lateral sclerosis, but they may not occur in cases with dominant lower motor neuron symptoms, including progressive muscle atrophy (Hobson et al., 2016).

3.2.9.2 Cortical Changes

- **Cortical hyperexcitability:** MND is linked to cortical hyperexcitability, which serves as an early indicator of upper motor neuron (UMN) impairment, as evidenced by transcranial magnetic stimulation (TMS) techniques (Vucic et al., 2008). To identify the heterogeneity of TMS results, motor-evoked units in patients with MND are found to be rather heterogeneous compared to the more homogeneous reactions found in healthy individuals. The changes observed in the corticomotoneuron system might be the reason behind the instability observed in the gait rhythm regulation in patients with MND (Menon et al., 2015).

- **Mirror neuron system:** One possible mechanism that might influence gait in MND is the mirror neuronal system (MNS), which is a multi-purpose cortical network involved in the imagined execution of motor action and in motor imagery. The system plays a critical role in controlling motor functions and the acquisition of skills related to gait and balance (Eisen et al., 2015). Studies that have been carried out using functional magnetic resonance imaging (fMRI) and other neurophysiological techniques have led to the discovery of the MNS in human beings. Such a system includes the frontoparietal mirror neuron network, which is a complex of several cortexes, such as the premotor, primary motor, primary somatosensory, supplementary motor, and inferior parietal lobules. The application of the fMRI technique in the context of motor imaging of the gait has shown an increase in the activity of the right prefrontal cortex and cerebellum (Allali et al., 2014). The mirror neuron system has been studied, and its implications on gait impairment have been examined among different neurodegenerative conditions, especially in patients who have suffered a stroke. One study found that stroke patients undergoing a program of observation and motor imagery training, which has been associated with the activity of the mirror neuron, showed more improvement in balance and gait as compared to those who were provided with traditional therapy (Kim & Lee, 2013). The value of the MNS can be justified by the studies of nonhuman primates who have discovered that the corticospinal neurons have mirror properties. The failure of this pathway in human beings, particularly in MND, may result in the malfunction of the MNS to regulate motor gait. Nevertheless, the precise contribution of the MNS to gait control in relation to the MND is not clearly known (Dharmadasa et al., 2018).

3.3 Conclusion

In general, it can be concluded that the diseases affecting the nervous system are the primary causes of disability since they directly influence the ability to control, coordinate, and regulate the processes in the body. In case of damage or dysfunction of the brain, spinal cord, and peripheral nerves, the impairment of mobility, touch, thinking, communication, and behaviour is crippling. Neurological conditions like Alzheimer's, Parkinson's, epilepsy, cerebral palsy, multiple sclerosis, motor neuron disease, Bell's palsy, and neurofibromatosis show the broad spectrum of clinical presentations of disorders of the nervous system. They also show the magnitude of the degree to which these conditions can be detrimental to normal body functions. They are usually progressive conditions, and with complicated complications that not only present a challenge to the healthcare but also to the families about long-term care needs. Besides medical consequences, patients with nervous system conditions are frequently confronted with the ruthless social and environmental barriers, including a lack of mobility, education and employment opportunities, and social isolation. These challenges have severe effects on the psychological well-being and the quality of life, in addition to causing emotional, physical, and financial stress to the care providers. Therefore, the therapies for the disorders of the nervous system should be conducted in a multidisciplinary fashion that involves early diagnosis, therapy, medical therapy, rehabilitation, and psychosocial support.

References

Adamson, P. M., Datta, K., Watkins, R., Recht, L. D., Hurd, R. E., & Spielman, D. M. (2024). Deuterium metabolic imaging for 3D mapping of glucose metabolism in humans with central nervous system lesions at 3T. *Magnetic Resonance in Medicine, 91*(1), 39–50.

Adlimoghaddam, A., Neuendorff, M., Roy, B., & Albensi, B. C. (2018). A review of clinical treatment considerations of donepezil in severe Alzheimer's disease. *CNS Neuroscience & Therapeutics, 24*(10), 876–888.

Ahmad, F. B. (2021). Provisional mortality data—United States, 2020. *MMWR: Morbidity and Mortality Weekly Report*, 70.

Ailon, T., Beauchamp, R., Miller, S., Mortenson, P., Kerr, J. M., Hengel, A. R., & Steinbok, P. (2015). Long-term outcome after selective dorsal rhizotomy in children with spastic cerebral palsy. *Child's Nervous System, 31*(3), 415–423.

Al-Chalabi, A., Van Den Berg, L. H., & Veldink, J. (2017). Gene discovery in amyotrophic lateral sclerosis: implications for clinical management. *Nature Reviews Neurology, 13*(2), 96–104.

Allali, G., van der Meulen, M., Beauchet, O., Rieger, S. W., Vuilleumier, P., & Assal, F. (2014). The neural basis of age-related changes in motor imagery of gait: an fMRI study. *Journals of Gerontology Series A: Biomedical Sciences and Medical Sciences, 69*(11), 1389–1398.

Alqahtani, S. Y., Almalki, Z. A., Alnafie, J. A., Alnemari, F. S., AlGhamdi, T. M., AlGhamdi, D. A., Albogami, L. O., & Ibrahim, M. (2024). Recurrent Bell's palsy: A comprehensive analysis of associated factors and outcomes. *Ear, Nose & Throat Journal*, 01455613241301230.

Anjum, I., Fayyaz, M., Wajid, A., Sohail, W., & Ali, A. (2018). Does obesity increase the risk of dementia: A literature review. *Cureus, 10*(5).

Antczak, J., Rusin, G., & Słowik, A. (2021). Transcranial magnetic stimulation as a diagnostic and therapeutic tool in various types of dementia. *Journal of Clinical Medicine, 10*(13), 2875.

Aravamuthan, B. R., Fehlings, D., Shetty, S., Fahey, M., Gilbert, L., Tilton, A., & Kruer, M. C. (2021). Variability in cerebral palsy diagnosis. *Pediatrics, 147*(2), e2020010066.

Armentano, M., Lucchino, L., Alisi, L., Chicca, A. V., Di Martino, V., Miraglia, E., Iannetti, L., Comberiati, A. M., Giustini, S., Lambiase, A., & Moramarco, A. (2023). Ophthalmic manifestation in neurofibromatosis type 2. *Applied Sciences, 13*(10), 6304.

Ascoli, M., Mastroianni, G., Gasparini, S., Striano, P., Cianci, V., Neri, S., Bova, V., Mammì, A., Gambardella, A., Labate, A., & Aguglia, U. (2021). Diagnostic and therapeutic approach to drug-resistant juvenile myoclonic epilepsy. *Expert Review of Neurotherapeutics, 21*(11), 1265–1273.

Babl, F. E., Herd, D., Borland, M. L., Kochar, A., Lawton, B., Hort, J., West, A., George, S., Zhang, M., Velusamy, K., & Sullivan, F. (2022). Efficacy of prednisolone for Bell palsy in children: A randomized, double-blind, placebo-controlled, multicenter trial. *Neurology, 99*(20), e2241–e2252.

Bachir, S., Shah, S., Shapiro, S., Koehler, A., Mahammedi, A., Samy, R. N., Zuccarello, M., Schorry, E., & Sengupta, S. (2021). Neurofibromatosis type 2 (NF2) and the implications for vestibular schwannoma and meningioma pathogenesis. *International Journal of Molecular Sciences, 22*(2), 690.

Banerjee, D., Das, P. K., & Mukherjee, J. (2023). Nervous system. In *Textbook of veterinary physiology* (pp. 265–293). Springer Nature.

Barbi, C., Pizzini, F. B., Tamburin, S., Martini, A., Pedrinolla, A., Laginestra, F. G., Giuriato, G., Martignon, C., Schena, F., & Venturelli, M. (2022). Brain structural and functional alterations in multiple sclerosis-related fatigue: A systematic review. *Neurology International, 14*(2), 506–535.

Beniczky, S., Tatum, W. O., Blumenfeld, H., Stefan, H., Mani, J., Maillard, L., Fahoum, F., Vinayan, K. P., Mayor, L. C., Vlachou, M., & Margitta, S. (2022). Seizure semiology: ILAE glossary of terms and their significance. *Epileptic Disorders, 3*, 447–495.

Bergen, D. C., & Silberberg, D. (2002). Nervous system disorders: A global epidemic. *Archives of Neurology, 59*(7), 1194–1196.

Beyrle, M., Sepp, D., Mühlau, M., Berthele, A., Engl, C., Hemmer, B., Zimmer, C., Wiestler, B., Kirschke, J. S., & Kertels, O. (2025). T1-weighted subtraction maps improve the detection of contrast-enhancing lesions in multiple sclerosis. *European Journal of Radiology*, 112576.

Bjornevik, K., Cortese, M., Healy, B. C., Kuhle, J., Mina, M. J., Leng, Y., Elledge, S. J., Niebuhr, D. W., Scher, A. I., Munger, K. L., & Ascherio, A. (2022). Longitudinal analysis reveals high prevalence of Epstein-Barr virus associated with multiple sclerosis. *Science, 375*(6578), 296–301.

Bottino, C. M., Castro, C. C., Gomes, R. L., Buchpiguel, C. A., Marchetti, R. L., & Neto, M. R. (2002). Volumetric MRI measurements can differentiate Alzheimer's disease, mild cognitive impairment, and normal aging. *International Psychogeriatrics, 14*(1), 59–72.

Boukhvalova, M. S., Mortensen, E., Lopez, D., Herold, B. C., & Blanco, J. C. (2022). Bell's palsy and lip HSV-1 infection: Importance of subcutaneous access. *Journal of Translational Science, 8*(1).

Branjerdporn, N., Ziviani, J., & Sakzewski, L. (2018). Goal-directed occupational therapy for children with unilateral cerebral palsy: Categorising and quantifying session content. *British Journal of Occupational Therapy, 81*(3), 138–146.

Bravo, G. Á., Foraster, A. C., Domínguez, D. L., & Vila, B. S. (2023). Movement disorders emergencies in metabolic disorders. In *International review of movement disorders* (Vol. 6, pp. 17–48). Academic Press.

Breijyeh, Z., & Karaman, R. (2020). Comprehensive review on Alzheimer's disease: Causes and treatment. *Molecules, 25*(24), 5789.

Caltabiano, R., Magro, G., Polizzi, A., Praticò, A. D., Ortensi, A., D'Orazi, V., Panunzi, A., Milone, P., Maiolino, L., Nicita, F., & Capone, G. L. (2017). A mosaic pattern of INI1/SMARCB1 protein expression distinguishes Schwannomatosis and NF2-associated peripheral schwannomas from solitary peripheral schwannomas and NF2-associated vestibular schwannomas. *Child's Nervous System, 33*(6), 933–940.

Cameron, M. H., & Nilsagard, Y. (2018). Balance, gait, and falls in multiple sclerosis. In *Handbook of clinical neurology* (Vol. 159, pp. 237–250).

Campian, J., & Gutmann, D. H. (2017). CNS tumors in neurofibromatosis. *Journal of Clinical Oncology, 35*(21), 2378–2385.

Canova, P. N., Charron, A. J., & Leib, D. A. (2024). Models of herpes simplex virus latency. *Viruses, 16*(5), 747.

Cao, Q., Cui, T., Wang, Q., Li, Z. M., Fan, S. H., Xiao, Z. M., Pan, S. Q., Zhou, Q., Lu, Z. N., & Shao, X. Q. (2022). The localization and lateralization of fear aura and its surgical prognostic value in patients with focal epilepsy. *Annals of Clinical and Translational Neurology, 9*(8), 1116–1124.

Carter, J. M., O'Hara, C., Dundas, G., Gilchrist, D., Collins, M. S., Eaton, K., Judkins, A. R., Biegel, J. A., & Folpe, A. L. (2012). Epithelioid malignant peripheral nerve sheath tumor arising in a schwannoma, in a patient with "neuroblastoma-like" schwannomatosis and a novel germline SMARCB1 mutation. *The American Journal of Surgical Pathology, 36*(1), 154–160.

Cerulli Irelli, E., Gesche, J., Schlabitz, S., Fortunato, F., Catania, C., Morano, A., Labate, A., Vorderwülbecke, B. J., Gambardella, A., Baykan, B., & Holtkamp, M. (2024). Epilepsy with generalized tonic–clonic seizures alone: Electroclinical features and prognostic patterns. *Epilepsia, 65*(1), 84–94.

Chen, J. L., Miller, D. T., Schmidt, L. S., Malkin, D., Korf, B. R., Eng, C., Kwiatkowski, D. J., & Giannikou, K. (2022). Mosaicism in tumor suppressor gene syndromes: Prevalence, diagnostic strategies, and transmission risk. *Annual Review of Genomics and Human Genetics, 23*(1), 331–361.

Cotsapas, C., & Mitrovic, M. (2018). Genome-wide association studies of multiple sclerosis. *Clinical & Translational Immunology, 7*(6), e1018.

Cox, B. C., Khattak, J. F., Starnes, K., Brinkmann, B. H., Tatum, W. O., Noe, K. H., Van Gompel, J. J., Miller, K. J., Marsh, W. R., Grewal, S. S., & Zimmerman, R. S. (2022). Subclinical seizures on stereotactic EEG: Characteristics and prognostic value. Seizure: European. *Journal of Epilepsy, 101*, 96–102.

Członkowska, A., Skowrońska, M., Dusek, P., Poujois, A., Laurencin, C., Litwin, T., Kurkowska-Jastrzębska, I., & Chabik, G. (2025). Modified unified Wilson's disease rating scale—scale presentation and pilot clinimetric testing. *Neurologia i Neurochirurgia Polska.*

Dal Bello, S., Martinuzzi, D., Tereshko, Y., Veritti, D., Sarao, V., Gigli, G. L., Lanzetta, P., & Valente, M. (2023). The present and future of optic pathway glioma therapy. *Cells, 12*(19), 2380.

de Castro, R. F., Crema, D., Neiva, F. C., Pinto, R. A., & Suzuki, F. A. (2022). Prevalence of herpes zoster virus reactivation in patients diagnosed with Bell's palsy. *The Journal of Laryngology & Otology, 136*(10), 975–978.

Dharmadasa, T., Matamala, J. M., Huynh, W., Zoing, M. C., & Kiernan, M. C. (2018). Motor neurone disease. In *Handbook of clinical neurology* (Vol. 159, pp. 345–357).

Dharmadasa, T., Matamala, J. M., & Kiernan, M. C. (2016). Treatment approaches in motor neurone disease. *Current Opinion in Neurology, 29*(5), 581–591.

Di Giacomo, R., Burini, A., Visani, E., Doniselli, F. M., Cuccarini, V., Garbelli, R., Marucci, G., De Santis, D., Didato, G., Deleo, F., & Pastori, C. (2023). Distinctive electro-clinical, neuroimaging and histopathological features of temporal encephaloceles associated to epilepsy. *Neurological Sciences, 44*(12), 4451–4463.

Douwes, J. P., van Eijk, R., Maas, S. L., Jansen, J. C., Aten, E., & Hensen, E. F. (2025). Genetic alterations in patients with NF2-related Schwannomatosis and sporadic vestibular Schwannomas. *Cancers, 17*(3), 393.

Eadie, M. J. (2024). WJ Adie and his "pyknolepsy," a century ago. *Journal of the History of the Neurosciences, 33*(2), 147–157.

Eaton, K. W., Tooke, L. S., Wainwright, L. M., Judkins, A. R., & Biegel, J. A. (2011). Spectrum of SMARCB1/INI1 mutations in familial and sporadic rhabdoid tumors. *Pediatric Blood & Cancer, 56*(1), 7–15.

Eisen, A., Lemon, R., Kiernan, M. C., Hornberger, M., & Turner, M. R. (2015). Does dysfunction of the mirror neuron system contribute to symptoms in amyotrophic lateral sclerosis? *Clinical Neurophysiology, 126*(7), 1288–1294.

Emanuele, M. J., Elia, A. E., Xu, Q., Thoma, C. R., Izhar, L., Leng, Y., Guo, A., Chen, Y. N., Rush, J., Hsu, P. W., & Yen, H. C. (2011). Global identification of modular cullin-RING ligase substrates. *Cell, 147*(2), 459–474.

Eminowicz, G. K., Raman, R., Conibear, J., & Plowman, P. N. (2012). Bevacizumab treatment for vestibular schwannomas in neurofibromatosis type two: Report of two cases, including responses after prior gamma knife and vascular endothelial growth factor inhibition therapy. *The Journal of Laryngology & Otology, 126*(1), 79–82.

Eroğlu, A., Güven, A. S., & Caksen, H. (2025). Juvenile myoclonic epilepsy adventure: A retrospective study. *Acta Neurologica Taiwanica, 34*(2), 76–80.

Esquenazi, A., Jost, W. H., Turkel, C. C., Wein, T., & Dimitrova, R. (2023). Treatment of adult spasticity with Botox (onabotulinumtoxinA): Development, insights, and impact. *Medicine, 102*(S1), e32376.

Esteves, A. D., Koyuncu, O. O., & Enquist, L. W. (2022). A pseudorabies virus serine/threonine kinase, US3, promotes retrograde transport in axons via Akt/mToRC1. *Journal of Virology, 96*(5), e01752–e01721.

Evans, D. G., Bowers, N. L., Tobi, S., Hartley, C., Wallace, A. J., King, A. T., Lloyd, S. K., Rutherford, S. A., Hammerbeck-Ward, C., Pathmanaban, O. N., & Freeman, S. R. (2018).

Schwannomatosis: A genetic and epidemiological study. *Journal of Neurology, Neurosurgery & Psychiatry, 89*(11), 1215–1219.

Evidente, V. G., Chrones, L., Revankar, R., Doshi, D., Abler, V., & Rashid, N. (2025). The modified functional status questionnaire for assessing patients with Parkinson's disease psychosis treated with Pimavanserin: A post hoc analysis. *Neurology and Therapy*, 1–1.

Eviston, T. J., Croxson, G. R., Kennedy, P. G., Hadlock, T., & Krishnan, A. V. (2015). Bell's palsy: Aetiology, clinical features and multidisciplinary care. *Journal of Neurology, Neurosurgery & Psychiatry, 86*(12), 1356–1361.

Fernández-Méndez, R., Wan, Y., Axon, P., & Joannides, A. (2023). Incidence and presentation of vestibular schwannoma: A 3-year cohort registry study. *Acta Neurochirurgica, 165*(10), 2903–2911.

Fissolo, N., Schaedelin, S., Villar, L. M., Lünemann, J. D., Correale, J., Rejdak, K., Schwab, N., Vilaseca, A., Held, F., García-Merino, A., & Bittner, S. (2025). Prognostic factors for multiple sclerosis symptoms in radiologically isolated syndrome. *JAMA Neurology*.

Forde, C., King, A. T., Rutherford, S. A., Hammerbeck-Ward, C., Lloyd, S. K., Freeman, S. R., Pathmanaban, O. N., Stapleton, E., Thomas, O. M., Laitt, R. D., & Stivaros, S. (2021). Disease course of neurofibromatosis type 2: A 30-year follow-up study of 353 patients seen at a single institution. *Neuro-Oncology, 23*(7), 1113–1124.

Frattini, V., Trifonov, V., Chan, J. M., Castano, A., Lia, M., Abate, F., Keir, S. T., Ji, A. X., Zoppoli, P., Niola, F., & Danussi, C. (2013). The integrated landscape of driver genomic alterations in glioblastoma. *Nature Genetics, 45*(10), 1141–1149.

Fu, W., Liang, J., Li, M., Song, G., Guo, J., Zheng, H., & Zhang, X. (2024). Effect of modified facial paralysis rehabilitation nursing on patients with facial paralysis after vestibular schwannoma surgery. *Heliyon, 10*(15).

Gajewski, B., Siger, M., Karlińska, I., Bednarski, I. A., Świderek-Matysiak, M., & Stasiołek, M. (2025). Brain atrophy and cognitive impairment in primary and secondary progressive multiple sclerosis cohort—similar progressive MS phenotype. *International Journal of Molecular Sciences, 26*(17), 8523.

Garzon, D. L., & Starr, N. B. (2023). Neurodivergence and behavioral and mental health disorders. *Burns' Pediatric Primary Care-E-Book*, 419.

Ghosh, R., Roy, D., Dubey, S., Das, S., & Benito-León, J. (2022). Movement disorders in multiple sclerosis: An update. *Tremor and Other Hyperkinetic Movements, 12*, 14.

Ghulaxe, Y., Joshi, A., Chavada, J., Huse, S., Kalbande, B., & Sarda, P. P. (2023). Understanding focal seizures in adults: A comprehensive review. *Cureus, 15*(11).

Gordon, A. M., Hung, Y. C., Brandao, M., Ferre, C. L., Kuo, H. C., Friel, K., Petra, E., Chinnan, A., & Charles, J. R. (2011). Bimanual training and constraint-induced movement therapy in children with hemiplegic cerebral palsy: A randomized trial. *Neurorehabilitation and Neural Repair, 25*(8), 692–702.

Goris, A., Vandebergh, M., McCauley, J. L., Saarela, J., & Cotsapas, C. (2022). Genetics of multiple sclerosis: Lessons from polygenicity. *The Lancet Neurology, 21*(9), 830–842.

Gouider, R., Souissi, A., Mrabet, S., Gharbi, A., Abida, Y., Kacem, I., & Gargouri-Berrechid, A. (2024). Environmental factors related to multiple sclerosis progression. *Journal of the Neurological Sciences, 464*, 123161.

Gourie-Devi, M., Nalini, A., & Sandhya, S. (2003). Early or late appearance of "dropped head syndrome" in amyotrophic lateral sclerosis. *Journal of Neurology, Neurosurgery & Psychiatry, 74*(5), 683–686.

Graves, J. S., Krysko, K. M., Hua, L. H., Absinta, M., Franklin, R. J., & Segal, B. M. (2023). Ageing and multiple sclerosis. *The Lancet Neurology, 22*(1), 66–77.

Haapasalo, A., & Hiltunen, M.. (n.d.) A report from the 8th Kuopio Alzheimer symposium.

Haddad, C., Sacre, H., Obeid, S., Salameh, P., & Hallit, S. (2021). Validation of the Arabic version of the "12-item short-form health survey"(SF-12) in a sample of Lebanese adults. *Archives of Public Health, 79*(1), 56.

Hadfield, K. D., Newman, W. G., Bowers, N. L., Wallace, A., Bolger, C., Colley, A., McCann, E., Trump, D., Prescott, T., & Evans, D. G. (2008). Molecular characterisation of SMARCB1 and NF2 in familial and sporadic schwannomatosis. *Journal of Medical Genetics, 45*(6), 332–339.

Hadfield, K. D., Smith, M. J., Urquhart, J. E., Wallace, A. J., Bowers, N. L., King, A. T., Rutherford, S. A., Trump, D., Newman, W. G., & Evans, D. G. (2010). Rates of loss of heterozygosity and mitotic recombination in NF2 schwannomas, sporadic vestibular schwannomas and schwannomatosis schwannomas. *Oncogene, 29*(47), 6216–6221.

Hakiminia, B., Alikiaii, B., Khorvash, F., & Mousavi, S. (2022). Oxidative stress and mitochondrial dysfunction following traumatic brain injury: From mechanistic view to targeted therapeutic opportunities. *Fundamental & Clinical Pharmacology, 36*(4), 612–662.

Hall, D. A., Shulman, J. M., Singleton, A., Ciga, S. B., Tosin, M. H., Ouyang, B., & Shulman, L. (2025). Racial disparities in Parkinson disease clinical phenotype, management, and genetics: Protocol for a prospective observational study. *JMIR Research Protocols, 14*(1), e60587.

Hampel, H., Mesulam, M. M., Cuello, A. C., Farlow, M. R., Giacobini, E., Grossberg, G. T., Khachaturian, A. S., Vergallo, A., Cavedo, E., Snyder, P. J., & Khachaturian, Z. S. (2018). The cholinergic system in the pathophysiology and treatment of Alzheimer's disease. *Brain, 141*(7), 1917–1933.

Harvey, S., & Shahwan, A. (2023). Typical absence seizures in children: Review with focus on EEG predictors of treatment response and outcome. *Seizure: European Journal of Epilepsy, 110*, 1.

Hausdorff, J. M., Lertratanakul, A., Cudkowicz, M. E., Peterson, A. L., Kaliton, D., & Goldberger, A. L. (2000). Dynamic markers of altered gait rhythm in amyotrophic lateral sclerosis. *Journal of Applied Physiology*.

Hausdorff, J. M., Mitchell, S. L., Firtion, R., Peng, C. K., Cudkowicz, M. E., Wei, J. Y., & Goldberger, A. L. (1997). Altered fractal dynamics of gait: reduced stride-interval correlations with aging and Huntington's disease. *Journal of Applied Physiology, 82*(1), 262–269.

Hill, C. S., & Loreto, A. (2021). Emergence of the Wallerian degeneration pathway as a mechanism of secondary brain injury. *Neural Regeneration Research, 16*(5), 980–981.

Hobson, E. V., Harwood, C. A., McDermott, C. J., & Shaw, P. J. (2016). Clinical aspects of motor neurone disease. *Medicine, 44*(9), 552–556.

Hoogmartens, J., Cacace, R., & Van Broeckhoven, C. (2021). Insight into the genetic etiology of Alzheimer's disease: A comprehensive review of the role of rare variants. *Alzheimer's & Dementia: Diagnosis, Assessment & Disease Monitoring, 13*(1), e12155.

Hsu H, Siwiec RM. StatPearls [Internet] StatPearls Publishing. . 2022.

Huser, A., Mo, M., & Hosseinzadeh, P. (2018). Hip surveillance in children with cerebral palsy. *Orthopedic Clinics, 49*(2), 181–190.

Ikeudenta, B. A., & Rutkofsky, I. H. (2020). Unmasking the enigma of cerebral palsy: A traditional review. *Cureus, 12*(10).

Jaafar, N., Bhatt, A., Eid, A., & Koubeissi, M. Z. (2022). The temporal lobe as a symptomatogenic zone in medial parietal lobe epilepsy. *Frontiers in Neurology, 13*, 804128.

Joint Committee on Infant Hearing, American Academy of Audiology, American Academy of Pediatrics, American Speech-Language-Hearing Association, Directors of Speech and Hearing Programs in State Health and Welfare Agencies. (2000). Year 2000 position statement: Principles and guidelines for early hearing detection and intervention programs. *Pediatrics, 106*(4), 798–817.

Kerr, W. T., Zhang, X., Hill, C. E., Janio, E. A., Chau, A. M., Braesch, C. T., Le, J. M., Hori, J. M., Patel, A. B., Allas, C. H., & Karimi, A. H. (2021). Epilepsy, dissociative seizures, and mixed: Associations with time to video-EEG. *Seizure*, (86), 116–122.

Khoury, R., Grysman, N., Gold, J., Patel, K., & Grossberg, G. T. (2018). The role of 5 HT6-receptor antagonists in Alzheimer's disease: An update. *Expert Opinion on Investigational Drugs, 27*(6), 523–533.

Kim, J. H., & Lee, B. H. (2013). Action observation training for functional activities after stroke: a pilot randomized controlled trial. *Neurorehabilitation, 33*(4), 565–574.

Kresak, J. L., & Walsh, M. (2016). Neurofibromatosis: A review of NF1, NF2, and schwannomatosis. *Journal of Pediatric Genetics, 5*(02), 098–104.

Kumar, A., Sidhu, J., Goyal, A., Tsao, J. W., & Doerr, C. (n.d.) Alzheimer disease (nursing).

Kurz, S. C., & Rogers, L. R. (2022). Vascular disorders: Epidemiology. In *Handbook of neuro-oncology neuroimaging* (Vol. 1, pp. 81–86). Academic Press.

Läderach, F., & Münz, C. (2022). Altered immune response to the Epstein–Barr virus as a prerequisite for multiple sclerosis. *Cells, 11*(17), 2757.

Landry, J. P., Schertz, K. L., Chiang, Y. J., Bhalla, A. D., Yi, M., Keung, E. Z., Scally, C. P., Feig, B. W., Hunt, K. K., Roland, C. L., & Guadagnolo, A. (2021). Comparison of cancer prevalence in patients with neurofibromatosis type 1 at an academic cancer center vs in the general population from 1985 to 2020. *JAMA Network Open, 4*(3), e210945.

Lee, J. M., Choi, Y. J., Yoo, M. C., & Yeo, S. G. (2023). Central facial nervous system biomolecules involved in peripheral facial nerve injury responses and potential therapeutic strategies. *Antioxidants, 12*(5), 1036.

Liljegren, M., Waldö, M. L., Rydbeck, R., & Englund, E. (2018). Police interactions among neuropathologically confirmed dementia patients: Prevalence and cause. *Alzheimer Disease & Associated Disorders, 32*(4), 346–350.

Lovras, M., Rana, A., Rani, S., Chauhan, A., Sridhar, S. B., Rajput, S., Malviya, R., & Wadhwa, T. (2025). Advancement in gene therapy for the treatment of Parkinson's disease: A comprehensive review. *Mini-Reviews in Medicinal Chemistry.*

Luna, E. B., Montovani, P. P., Rozza-de-Menezes, R. E., & Cunha, K. S. (2021). Neurofibromin expression by normal salivary glands. *Head & Face Medicine, 17*(1), 5.

Luo, B., & Ding, L. (2022). Ion channels and ions as therapeutic targets and strategies for herpes simplex virus infection. *Future Virology, 17*(11), 829–834.

Ly, N., Powers, B. R., & Chaiet, S. R. (2022). Adherence to clinical practice guidelines for treatment of Bell's palsy. *WMJ, 121*(4), 274–279.

Maccioni, R. B., González, A., Andrade, V., Cortés, N., Tapia, J. P., & Guzmán-Martínez, L. (2018). Alzheimer s disease in the perspective of neuroimmunology. *The Open Neurology Journal, 12*, 50.

Majounie, E., Renton, A. E., Mok, K., Dopper, E. G., Waite, A., Rollinson, S., Chiò, A., Restagno, G., Nicolaou, N., Simon-Sanchez, J., & Van Swieten, J. C. (2012). Frequency of the C9orf72 hexanucleotide repeat expansion in patients with amyotrophic lateral sclerosis and frontotemporal dementia: A cross-sectional study. *The Lancet Neurology, 11*(4), 323–330.

Malviya, R., & Rajput, S. (2025). *Artificial intelligence for neural health: Diagnosis and treatment.* CRC Press.

Marinelli, J. P., Schnurman, Z., Killeen, D. E., Nassiri, A. M., Hunter, J. B., Lees, K. A., Lohse, C. M., Roland, J. T., Jr., Golfinos, J. G., Kondziolka, D., & Link, M. J. (2022). Long-term natural history and patterns of sporadic vestibular schwannoma growth: A multi-institutional volumetric analysis of 952 patients. *Neuro-Oncology, 24*(8), 1298–1306.

Marques, A., Okpali, G., Liepshutz, K., & Ortega-Villa, A. M. (2022). Characteristics and outcome of facial nerve palsy from Lyme neuroborreliosis in the United States. *Annals of Clinical and Translational Neurology, 9*(1), 41–49.

Marrie, R. A., Fisk, J. D., Fitzgerald, K., Kowalec, K., Maxwell, C., Rotstein, D., Salter, A., & Tremlett, H. (2023). Etiology, effects and management of comorbidities in multiple sclerosis: Recent advances. *Frontiers in Immunology, 14*, 1197195.

Martin, A. R., Reddy, R., & Fehlings, M. G. (2011). Dropped head syndrome: diagnosis and management. *Evidence-Based Spine-Care Journal, 2*(2), 41–47.

Mazumder, R., Lagoro, D. K., Nariai, H., Danieli, A., Eliashiv, D., Engel, J., Jr., Dalla Bernardina, B., Kegele, J., Lerche, H., Sejvar, J., & Matuja, W. (2022). Ictal electroencephalographic characteristics of nodding syndrome: A comparative case-series from South Sudan, Tanzania, and Uganda. *Annals of Neurology, 92*(1), 75–80.

McGonigal, A. (2022). Frontal lobe seizures: Overview and update. *Journal of Neurology, 269*(6), 3363–3371.

Menchetti, I., McAllister, K., Walker, D., & Donnan, P. T. (2021). Surgical interventions for the early management of Bell's palsy. *Cochrane Database of Systematic Reviews, 1*.

Mendoza-Sengco, P., Chicoine, C. L., & Vargus-Adams, J. (2023). Early cerebral palsy detection and intervention. *Pediatric Clinics, 70*(3), 385–398.

Menon, P., Geevasinga, N., Yiannikas, C., Howells, J., Kiernan, M. C., & Vucic, S. (2015). Sensitivity and specificity of threshold tracking transcranial magnetic stimulation for diagnosis of amyotrophic lateral sclerosis: A prospective study. *The Lancet Neurology, 14*(5), 478–484.

Merker, V. L., Esparza, S., Smith, M. J., Stemmer-Rachamimov, A., & Plotkin, S. R. (2012). Clinical features of schwannomatosis: A retrospective analysis of 87 patients. *The Oncologist, 17*(10), 1317–1322.

Mistri, D., Cacciaguerra, L., Valsasina, P., Pagani, E., Filippi, M., & Rocca, M. A. (2023). Cognitive function in primary and secondary progressive multiple sclerosis: A multiparametric magnetic resonance imaging study. *European Journal of Neurology, 30*(9), 2801–2810.

Montalban, X., Lebrun-Frénay, C., Oh, J., Arrambide, G., Moccia, M., Amato, M. P., Amezcua, L., Banwell, B., Bar-Or, A., Barkhof, F., & Butzkueven, H. (2025). Diagnosis of multiple sclerosis: 2024 revisions of the McDonald criteria. *The Lancet Neurology, 24*(10), 850–865.

Montes, J., Cheng, B., Diamond, B., Doorish, C., Mitsumoto, H., & Gordon, P. H. (2007). The timed up and go test: Predicting falls in ALS. *Amyotrophic Lateral Sclerosis, 8*(5), 292–295.

Morgan, C., Fahey, M., Roy, B., & Novak, I. (2018). Diagnosing cerebral palsy in full-term infants. *Journal of Paediatrics and Child Health, 54*(10), 1159–1164.

Morris KA, Golding JF, Axon PR, Afridi S, Blesing C, Ferner RE, Halliday D, Jena R, Pretorius PM, UK NF2 Research group, Evans DG. Bevacizumab in neurofibromatosis type 2 (NF2) related vestibular schwannomas: A nationally coordinated approach to delivery and prospective evaluation. Neuro-Oncology Practice 2016;3(4):281–289.

Mukhopadhyay, S., Maitra, A., & Choudhury, S. (2021). Selumetinib: The first ever approved drug for neurofibromatosis-1 related inoperable plexiform neurofibroma. *Current Medical Research and Opinion, 37*(5), 789–794.

Murray, C. J. (2022). The global burden of disease study at 30 years. *Nature Medicine, 28*(10), 2019–2026.

Murray, C. J. (2024). Findings from the global burden of disease study 2021. *The Lancet, 403*(10440), 2259–2262.

Myers, M. G., Jr., Affinati, A. H., Richardson, N., & Schwartz, M. W. (2021). Central nervous system regulation of organismal energy and glucose homeostasis. *Nature Metabolism, 3*(6), 737–750.

Napolitano, F., Dell'Aquila, M., Terracciano, C., Franzese, G., Gentile, M. T., Piluso, G., Santoro, C., Colavito, D., Patanè, A., De Blasiis, P., & Sampaolo, S. (2022). Genotype-phenotype correlations in neurofibromatosis type 1: Identification of novel and recurrent NF1 gene variants and correlations with neurocognitive phenotype. *Genes, 13*(7), 1130.

Naseri, A., Nasiri, E., Sahraian, M. A., Daneshvar, S., & Talebi, M. (2021). Clinical features of late-onset multiple sclerosis: A systematic review and meta-analysis. *Multiple Sclerosis and Related Disorders, 50*, 102816.

Neckel, N., Nahles, S., Heiland, M., Audebert, H., Zdunczyk, A., Guntinas-Lichius, O., Preissner, R., & Preissner, S. (2025). Risk factors associated with Bell's palsy: A real-world analysis of 281,600 patients. *European Journal of Neurology, 32*(8), e70336.

Neff, B. A., Wiet, R. M., Lasak, J. M., Cohen, N. L., Pillsbury, H. C., Ramsden, R. T., & Welling, D. B. (2007). Cochlear implantation in the neurofibromatosis type 2 patient: Long-term follow-up. *The Laryngoscope, 117*(6), 1069–1072.

Neville, C., Gwynn, T., Young, K., Jordan, E., Malhotra, R., Nduka, C., & Kannan, R. Y. (2022). Comparative study of multimodal therapy in facial palsy patients. *Archives of Plastic Surgery, 49*(05), 633–641.

Nicolas, G., Acuña-Hidalgo, R., Keogh, M. J., Quenez, O., Steehouwer, M., Lelieveld, S., Rousseau, S., Richard, A. C., Oud, M. S., Marguet, F., & Laquerrière, A. (2018). Somatic variants in autosomal dominant genes are a rare cause of sporadic Alzheimer's disease. *Alzheimer's & Dementia, 14*(12), 1632–1639.

Nordberg, J., Schaper, F. L., Bucci, M., Nummenmaa, L., & Joutsa, J. (2023). Brain lesion locations associated with secondary seizure generalization in tumors and strokes. *Human Brain Mapping, 44*(8), 3136–3146.

Noteboom, S., Seiler, M., Chien, C., Rane, R. P., Barkhof, F., Strijbis, E. M., Paul, F., Schoonheim, M. M., & Ritter, K. (2024). Evaluation of machine learning-based classification of clinical impairment and prediction of clinical worsening in multiple sclerosis. *Journal of Neurology, 271*(8), 5577–5589.

Novak, I. (2014). Evidence-based diagnosis, health care, and rehabilitation for children with cerebral palsy. *Journal of Child Neurology, 29*(8), 1141–1156.

Novak, I., Cusick, A., & Lannin, N. (2009). Occupational therapy home programs for cerebral palsy: Double-blind, randomized, controlled trial. *Pediatrics, 124*(4), e606–e614.

Orrell, R. W., & Guiloff, R. J. (2023). Clinical aspects of motor neurone disease. *Medicine, 51*(9), 658–662.

O'shea, T. M. (2008). Diagnosis, treatment, and prevention of cerebral palsy. *Clinical Obstetrics and Gynecology, 51*(4), 816–828.

Ou, Z., Pan, J., Tang, S., Duan, D., Yu, D., Nong, H., & Wang, Z. (2021). Global trends in the incidence, prevalence, and years lived with disability of Parkinson's disease in 204 countries/territories from 1990 to 2019. *Frontiers in Public Health, 9*, 776847.

Ouerdani, A., Goutagny, S., Kalamarides, M., Trocóniz, I. F., & Ribba, B. (2016). Mechanism-based modeling of the clinical effects of bevacizumab and everolimus on vestibular schwannomas of patients with neurofibromatosis type 2. *Cancer Chemotherapy and Pharmacology, 77*(6), 1263–1273.

Park, K. S. (2023). Nervous system. In *Humans and electricity: Understanding body electricity and applications* (Vol. 4, pp. 27–51). Springer.

Paroni, G., Bisceglia, P., & Seripa, D. (2019). Understanding the amyloid hypothesis in Alzheimer's disease. *Journal of Alzheimer's Disease, 68*(2), 493–510.

Pavone, V., Testa, G., Restivo, D. A., Cannavò, L., Condorelli, G., Portinaro, N. M., & Sessa, G. (2016). Botulinum toxin treatment for limb spasticity in childhood cerebral palsy. *Frontiers in Pharmacology, 7*, 29.

Peduto, C., Zanobio, M., Nigro, V., Perrotta, S., Piluso, G., & Santoro, C. (2023). Neurofibromatosis type 1: Pediatric aspects and review of genotype–phenotype correlations. *Cancers, 15*(4), 1217.

Petersen, R. C. (2018). How early can we diagnose Alzheimer disease (and is it sufficient)? The 2017 Wartenberg lecture. *Neurology, 91*(9), 395–402.

Pialat, P. M., Fieux, M., Tringali, S., Beldjoudi, G., Pommier, P., & Tanguy, R. (2021). Vestibular schwannoma: Results of hypofractionated stereotactic radiation therapy. *Advances in Radiation Oncology, 6*(4), 100694.

Plotkin, S. R., Halpin, C., McKenna, M. J., Loeffler, J. S., Batchelor, T. T., & Barker, F. G. (2010). Erlotinib for progressive vestibular schwannoma in neurofibromatosis 2 patients. *Otology & Neurotology, 31*(7), 1135–1143.

Porsteinsson, A. P., Isaacson, R. S., Knox, S., Sabbagh, M. N., & Rubino, I. (2021). Diagnosis of early Alzheimer's disease: Clinical practice in 2021. *The Journal of Prevention of Alzheimer's Disease, 8*(3), 371–386.

Prabhu, P., Aryal, S., Rajan, A., & Nisha, K. V. (2025). Exploring the impact of Misophonia through the lens of the World Health Organization's international classification of functioning, disability and health framework. *Journal of the American Academy of Audiology., 36*(4), 321.

Qiu, C., Kivipelto, M., & Von Strauss, E. (2009). Epidemiology of Alzheimer's disease: Occurrence, determinants, and strategies toward intervention. *Dialogues in Clinical Neuroscience, 11*(2), 111–128.

Radovanović, S., Milićev, M., Perić, S., Basta, I., Kostić, V., & Stević, Z. (2014). Gait in amyotrophic lateral sclerosis: Is gait pattern differently affected in spinal and bulbar onset of the disease during dual task walking? *Amyotrophic Lateral Sclerosis and Frontotemporal Degeneration, 15*(7–8), 488–493.

Rajput, S., Malviya, R., Bahadur, S., & Puri, D. (2023). Recent updates on the development of therapeutics for the targeted treatment of Alzheimer's disease. *Current Pharmaceutical Design, 29*(35), 2802–2813.

Rana, A., Malviya, R., Rajput, S., Sridhar, S. B., & Wadhwa, T. (2025). Trends in nanoparticle-based strategies for the management of neuroinflammation. *CNS and Neurological Disorders: Drug Targets*.

Rashad, A., Rasool, A., Shaheryar, M., Sarfraz, A., Sarfraz, Z., Robles-Velasco, K., & Cherrez-Ojeda, I. (2022). Donanemab for Alzheimer's disease: A systematic review of clinical trials. In *Healthcare* (Vol. 11, p. 32). MDPI.

Rider, F., Turchinets, A., Druzhkova, T., Kustov, G., Guekht, A., & Gulyaeva, N. (2024). Dissimilar changes in serum cortisol after epileptic and psychogenic non-epileptic seizures: A promising biomarker in the differential diagnosis of paroxysmal events? *International Journal of Molecular Sciences, 25*(13), 7387.

Rousseau, G., Noguchi, T., Bourdon, V., Sobol, H., & Olschwang, S. (2011). SMARCB1/INI1 germline mutations contribute to 10% of sporadic schwannomatosis. *BMC Neurology, 11*(1), 9.

Rovira, À., & Auger, C. (2021). Beyond McDonald: Updated perspectives on MRI diagnosis of multiple sclerosis. *Expert Review of Neurotherapeutics, 21*(8), 895–911.

Ruggieri, M., & Huson, S. M. (2001). The clinical and diagnostic implications mosaicism in the neurofibromatoses. *Neurology, 56*(11), 1433–1443.

Sadler, K. V., Bowers, N. L., Hartley, C., Smith, P. T., Tobi, S., Wallace, A. J., King, A., Lloyd, S. K., Rutherford, S., Pathmanaban, O. N., & Hammerbeck-Ward, C. (2021). Sporadic vestibular schwannoma: A molecular testing summary. *Journal of Medical Genetics, 58*(4), 227–233.

Saifee, T., Farmer, S., Shah, S., & Choi, D. (2024). Spinal column and spinal cord disorders. In *Neurology: A queen square textbook* (pp. 463–498).

Salehi, Z., Naghizadeh, M. M., Ezabadi, S. G., Ebrahimitirtashi, A., Kasbi, N. A., Khodaie, F., Aliyari, S., Ashtari, F., Baghbanian, S. M., Nabavi, S. M., & Hosseini, S. (2024). Consanguineous marriage among familial multiple sclerosis subjects: A national registry-based study. *Heliyon, 10*(12).

Salehi, A., Wesson Ashford, J., Mufson, J., & E. (2016). Editorial (thematic issue: The link between alzheimer's disease and Down syndrome. A historical perspective). *Current Alzheimer Research, 13*(1), 2–6.

Santos, C. Y., Snyder, P. J., Wu, W. C., Zhang, M., Echeverria, A., & Alber, J. (2017). Pathophysiologic relationship between Alzheimer's disease, cerebrovascular disease, and cardiovascular risk: A review and synthesis. *Alzheimer's & Dementia: Diagnosis, Assessment & Disease Monitoring, 7*, 69–87.

Saraniti, C., & Verro, B. (2024). Reanimation techniques of peripheral facial paralysis: A comprehensive review focusing on surgical and bioengineering approaches. *Journal of Clinical Medicine, 13*(20), 6124.

Scheffzek, K., Ahmadian, M. R., Wiesmüller, L., Kabsch, W., Stege, P., Schmitz, F., & Wittinghofer, A. (1998). Structural analysis of the GAP-related domain from neurofibromin and its implications. *The EMBO Journal*.

Schwab, R., England, A., Schwab, Z., England, A., Schwab, R. Projection technique for evaluating surgery in Parkinson's disease. 1969. Royal College of Surgeons in Edinburgh: E. & S. Livingstone Ltd.[Google Scholar]. 2023.

Shi, Z., Ma, L., Luo, K., Bajaj, M., Chawla, S., Natarajan, G., Hagberg, H., & Tan, S. (2017). Chorioamnionitis in the development of cerebral palsy: A meta-analysis and systematic review. *Pediatrics, 139*(6).

Shikara, M., Bridgham, K., Ludeman, E., Vakharia, K., & Justicz, N. (2023). Selective neurectomy for treatment of post-facial paralysis synkinesis: A systematic review. *Facial Plastic Surgery., 39*(02), 190–200.

Shirai, R., & Yamauchi, J. (2022). New insights into risk genes and their candidates in multiple sclerosis. *Neurology International, 15*(1), 24–39.

Shorvon, S., Diehl, B., Duncan, J., Koepp, M., Rugg-Gunn, F., Sander, J., Walker, M., & Wehner, T. (2016). Epilepsy and related disorders. In *Neurology: A queen square textbook* (pp. 221–287).

Sialakis, C., Frantzana, A., Iliadis, C., Ouzounakis, P., & Kourkouta, L. (2025). Effectiveness of steroids and antiviral agents in the treatment of Bell's palsy. *Medicine and Pharmacy Reports, 98*(3), 267.

Singh, A., Deshmukh, P., & Deshmukh, P. T. (2022). Bell's palsy: A review. *Cureus, 14*(10).

Sirven, J. I. (2002). Classifying seizures and epilepsy: A synopsis. In Seminars in neurology. (Vol. 22, No. 03, pp. 237–246). Copyright© 2002 by Thieme Medical Publishers, Inc., 333 Seventh Avenue, New York, NY 10001, USA.

Smith, M. J., Isidor, B., Beetz, C., Williams, S. G., Bhaskar, S. S., Richer, W., O'Sullivan, J., Anderson, B., Daly, S. B., Urquhart, J. E., & Fryer, A. (2015). Mutations in LZTR1 add to the complex heterogeneity of schwannomatosis. *Neurology, 84*(2), 141–147.

Sobański, M., Zacharzewska-Gondek, A., Waliszewska-Prosół, M., Sąsiadek, M. J., Zimny, A., & Bladowska, J. (2020). A review of neuroimaging in rare neurodegenerative diseases. *Dementia and Geriatric Cognitive Disorders, 49*(6), 544–556.

Soldan, S. S., & Lieberman, P. M. (2023). Epstein–Barr virus and multiple sclerosis. *Nature Reviews Microbiology, 21*(1), 51–64.

Specchio, N., Wirrell, E. C., Scheffer, I. E., Nabbout, R., Riney, K., Samia, P., Guerreiro, M., Gwer, S., Zuberi, S. M., Wilmshurst, J. M., & Yozawitz, E. (2022). International league against epilepsy classification and definition of epilepsy syndromes with onset in childhood: Position paper by the ILAE task force on nosology and definitions. *Epilepsia, 63*(6), 1398–1442.

Stephens, K., Kayes, L., Riccardi, V. M., Rising, M., Sybert, V. P., & Pagon, R. A. (1992). Preferential mutation of the neurofibromatosis type 1 gene in paternally derived chromosomes. *Human Genetics, 88*(3), 279–282.

Steriade, C., Sperling, M. R., DiVentura, B., Lozano, M., Shellhaas, R. A., Kessler, S. K., Dlugos, D., & French, J. (2022). Proposal for an updated seizure classification framework in clinical trials. *Epilepsia, 63*(3), 565–572.

Sugavaneswaran, L., Umapathy, K., & Krishnan, S. (2012). Ambiguity domain-based identification of altered gait pattern in ALS disorder. *Journal of Neural Engineering, 9*(4), 046004.

Swash, M. (2012). Why are upper motor neuron signs difficult to elicit in amyotrophic lateral sclerosis? *Journal of Neurology, Neurosurgery & Psychiatry, 83*(6), 659–662.

Tamura, R. (2021). Current understanding of neurofibromatosis type 1, 2, and schwannomatosis. *International Journal of Molecular Sciences, 22*(11), 5850.

Tamura, R., Fujioka, M., Morimoto, Y., Ohara, K., Kosugi, K., Oishi, Y., Sato, M., Ueda, R., Fujiwara, H., Hikichi, T., & Noji, S. (2019). A VEGF receptor vaccine demonstrates preliminary efficacy in neurofibromatosis type 2. *Nature Communications, 10*(1), 5758.

Tawfik, H. A., & Dutton, J. J. (2023). Facial nerve palsy and the eye: Etiology, diagnosis, and management. *International Ophthalmology Clinics, 63*(3), 75–94.

Tedeschi, R. (2024). Mapping the current research on mindfulness interventions for individuals with cerebral palsy: A scoping review. *Neuropediatrics, 55*(02), 077–082.

Teixeira, L. (2024). The nervous system and associated disorders. *British Journal of Nursing, 33*(4), 194–199.

Terwilliger, J. D. (2022). Gene mapping and human disease. In *Genetics of substance use: Research and clinical aspects* (pp. 147–175). Springer.

Therriault, J., Zimmer, E. R., Benedet, A. L., Pascoal, T. A., Gauthier, S., & Rosa-Neto, P. (2022). Staging of Alzheimer's disease: Past, present, and future perspectives. *Trends in Molecular Medicine, 28*(9), 726–741.

Thompson, A. J., Baranzini, S. E., Geurts, J., Hemmer, B., & Ciccarelli, O. (2018). Seminar multiple sclerosis. *Lancet, 391*, 1622–1636.

Tong, S., Devine, W. P., & Shieh, J. T. (2022). Tumor and constitutional sequencing for neurofibromatosis type 1. *JCO Precision Oncology, 6*, e2100540.

Tosi, U., Maayan, O., An, A., Lavieri, M. E., Guadix, S. W., DeRosa, A. P., Christos, P. J., Pannullo, S., Stieg, P. E., Brandmaier, A., & Knisely, J. P. (2022). Stereotactic radiosurgery for vestibular schwannomas in neurofibromatosis type 2 patients: A systematic review and meta-analysis. *Journal of Neuro-Oncology, 156*(2), 431–441.

Traxinger, K., Kelly, C., Johnson, B. A., Lyles, R. H., & Glass, J. D. (2013). Prognosis and epidemiology of amyotrophic lateral sclerosis: Analysis of a clinic population, 1997–2011. *Neurology: Clinical Practice, 3*(4), 313–320.

Tustin, K., & Patel, A. (2017). A critical evaluation of the updated evidence for casting for equinus deformity in children with cerebral palsy. *Physiotherapy Research International, 22*(1), e1646.

van den Munckhof, P., Christiaans, I., Kenter, S. B., Baas, F., & Hulsebos, T. J. (2012). Germline SMARCB1 mutation predisposes to multiple meningiomas and schwannomas with preferential location of cranial meningiomas at the falx cerebri. *Neurogenetics, 13*(1), 1–7.

Velnar, T., Kocivnik, N., & Bosnjak, R. (2023). Clinical infections in neurosurgical oncology: An overview. *World Journal of Clinical Cases, 11*(15), 3418.

Vieluf, S., Hasija, T., Schreier, P. J., El Atrache, R., Hammond, S., Touserkani, F. M., Sarkis, R. A., Loddenkemper, T., & Reinsberger, C. (2021). Generalized tonic-clonic seizures are accompanied by changes of interrelations within the autonomic nervous system. *Epilepsy & Behavior, 124*, 108321.

Vitrikas, K., Dalton, H., & Breish, D. (2020). Cerebral palsy: An overview. *American Family Physician, 101*(4), 213–220.

Vucic, S., Cheah, B. C., & Kiernan, M. C. (2011). Maladaptation of cortical circuits underlies fatigue and weakness in ALS. *Amyotrophic Lateral Sclerosis, 12*(6), 414–420.

Vucic, S., Nicholson, G. A., & Kiernan, M. C. (2008). Cortical hyperexcitability may precede the onset of familial amyotrophic lateral sclerosis. *Brain, 131*(6), 1540–1550.

Wang, X., Du, X. G., Teh, S. H., & Wang, X. H. (2025). Risk factors for spastic cerebral palsy: A retrospective cross-sectional study and literature review. *Italian Journal of Pediatrics, 51*(1), 250.

Wang, H. E., Scholly, J., Triebkorn, P., Sip, V., Villalon, S. M., Woodman, M. M., Le Troter, A., Guye, M., Bartolomei, F., & Jirsa, V. (2021). VEP atlas: An anatomic and functional human brain atlas dedicated to epilepsy patients. *Journal of Neuroscience Methods, 348*, 108983.

Wang, Y. Q., Wen, Y., Wang, M. M., Zhang, Y. W., & Fang, Z. X. (2021). Prolactin levels as a criterion to differentiate between psychogenic non-epileptic seizures and epileptic seizures: A systematic review. *Epilepsy Research, 169*, 106508.

Wang, A., Xie, W., & Zhang, J. (2025). The synergistic role of viral infection and immune response in the pathogenesis of facial palsy. *Journal of NeuroVirology*, 1–1.

Wentworth, S., Pinn, M., Bourland, J. D., Deguzman, A. F., Ekstrand, K., Ellis, T. L., Glazier, S. S., McMullen, K. P., Munley, M., Stieber, V. W., & Tatter, S. B. (2009). Clinical experience with radiation therapy in the management of neurofibromatosis-associated central nervous system tumors. *International Journal of Radiation Oncology, Biology, Physics, 73*(1), 208–213.

Wergeland, S., Mannseth, J., Atula, S., Forsberg, L., Glaser, A., Joensen, H., Magyari, M., Soilu-Hanninen, M., Viitala, M., Hillert, J., & Grytten, N. (2023). Long-term survival in multiple sclerosis is improving (P5–3.011). *Neurology, 100*(17_supplement_2), 1834.

Wood, C., Owen, S., Ebden, S., Anand, B., Wardle, M., Hamandi, K., Kreft, K. L., Robertson, N. P., & Tallantyre, E. C. (2025). Multiple sclerosis and seizures: Clinical, diagnostic and therapeutic correlations. *Brain and Behavior, 15*(5), e70511.

World Health Organization. (2023). Trends in maternal mortality 2000 to 2020: Estimates by WHO, UNICEF, UNFPA, World Bank Group and UNDESA/Population Division. World Health Organization.

Xiang, Y., Zhang, M., Jiang, D., Su, Q., & Shi, J. (2023). The role of inflammation in autoimmune disease: A therapeutic target. *Frontiers in Immunology, 14*, 1267091.

Yokoi, K., Kawasaki, I., Takeda, A., Eakman, A. M., & Hirayama, K. (2025). Profile of Independence in activities of daily living among patients with Parkinson's disease: A retrospective observational study. *The American Journal of Occupational Therapy, 79*(3), 7903205040.

Younger, D. S. (2023). Multiple sclerosis: Motor dysfunction. *Handbook of Clinical Neurology, 196*, 119–147.

Zawar, I., Shreshtha, B., Benech, D., Burgess, R. C., Bulacio, J., & Knight, E. M. (2024). Electrographic features of epilepsy with eyelid Myoclonia with Photoparoxysmal responses. *Journal of Clinical Neurophysiology, 41*(1), 83–92.

Zetterberg, H., & Bendlin, B. B. (2021). Biomarkers for Alzheimer's disease—preparing for a new era of disease-modifying therapies. *Molecular Psychiatry, 26*(1), 296–308.

Zhang, J., Cortese, R., De Stefano, N., & Giorgio, A. (2021). Structural and functional connectivity substrates of cognitive impairment in multiple sclerosis. *Frontiers in Neurology, 12*, 671894.

Zhu, Y., Liang, T., Liu, S., Xu, J., Zhao, H., Wang, Y., Qiu, J., Liu, D., & Sun, Y. (2025). Comparison of glucocorticoids combined with antiviral drugs versus glucocorticoids alone in the management of Bell's Palsy: A systematic review and meta-analysis of randomized clinical trials. *American Journal of Otolaryngology, 46*(1), 104583.

Zoila, F., de Stefano, M. I., Sgobbio, A., Panaro, M. A., Maffione, A. B., Antonucci, L., Benameur, T., Massaro, M., Frota Gaban, S. V., Filannino, F. M., & Porro, C. (2025). Yoga for neurodegenerative disorders: Therapeutic effects, mechanisms, and applications in Alzheimer's and Parkinson's disease. *Sports, 13*(12), 458.

Chapter 4
Muscle and Motor System Disability

Abstract The disabilities discussed in this chapter include motor and muscular impairments. It analyses that muscle weakness and lack of motor control and coordination may affect movement and daily activities. In addition, the chapter looks into the root causes of motor impairments and the way they impact physical functions and locomotion. It highlights the significance of early assessment and rehabilitation procedures to improve motor skills and the quality of life in general. Another important aspect in this chapter is the fact that people who experience a disability of the muscle and motor system, such as muscular dystrophy, spinal muscular atrophy, and cerebral palsy, usually experience social issues like inaccessibility of open space, reliance on the caregiver, physical activity exclusion, and limited opportunities that affect their personal and family life.

Keywords Motor disability · Neuromuscular disorders · Motor impairment · Muscle weakness · Rehabilitation · Mobility limitation · Assistive devices

4.1 Introduction

Motor system disabilities are an example of progressive neurological disorders characterized by the degeneration of motor neurons, which play a vital role in controlling skeletal muscle movement, such as locomotion and respiration. The main types of motor neuron illnesses are muscular dystrophy, amyotrophic lateral sclerosis, spinal muscular atrophy, myasthenia gravis, peripheral neuropathy, myopathy, spinal cord injury, and post-polio syndrome (Feldman et al., 2021). The upper neuron to lower neuron communication in the brain to the lower neurons in the spinal cord is the key to muscle functioning; the impairment of this communication leads to the weakness of muscles, muscle atrophy, and fibrils. Damage to the lower motor neurons could lead to muscle rigidity and reflexes, which hinder voluntary movement. Motor system disabilities are classified based on their aetiology, which may be hereditary (i.e. genetic mutations) or sporadic (i.e. not inherited by a family

R. Malviya, S. Rajput, *Neurogenetic and Neurodevelopmental Disabilities*, SpringerBriefs in Modern Perspectives on Disability Research,
https://doi.org/10.1007/978-981-95-9223-4_4

member), as well as by the motor neurons they affect (Mathis et al., 2024). All motor system disability has the consequence of progressive muscular weakness and possible physical disability, with some causing death, especially where respiratory muscles are affected, causing respiratory failure. The symptoms are usually dyspnoea, orthopnoea, frequent pulmonary infections, sleep disorders, cognitive impairments, disorientation, headaches in the morning, and fatigue (Singh, 2020). Figure 4.1 and Table 4.1 illustrate various muscle and motor system disabilities.

Movement disorders also encompass several neurological diseases that affect the control of movements of an individual, either an increase or a decrease in motor activity. Some movements can be termed as voluntary, which can be made by conscious control, and involuntary, which takes place without conscious control. There are several types of movement disorders, and each has different symptoms: dystonia is the presence of involuntary muscle contractions that lead to abnormal twisting movements; chorea is the occurrence of brief and rapid involuntary movements; and Parkinsonism is the manifestation of bradykinesia, stiffness, tremors, and loss of balance (Abdo et al., 2010). These conditions can be treated using management plans such as pharmaceutical therapy, therapeutic management, or surgery. Also, it would be possible to manage any underlying conditions that contribute to the worsening of the movement disorder, which would potentially relieve symptoms. Motor system disability mostly occurs in both adults and children, though due to various causes (Trabacca et al., 2024). In children, such disorders are usually associated with genetic anomalies, as is seen with spinal muscular atrophy when the symptoms present themselves during birth or during early infancy. Adult motor neuron disorders, on the contrary, are sporadic in nature and mostly occur after age 50, although

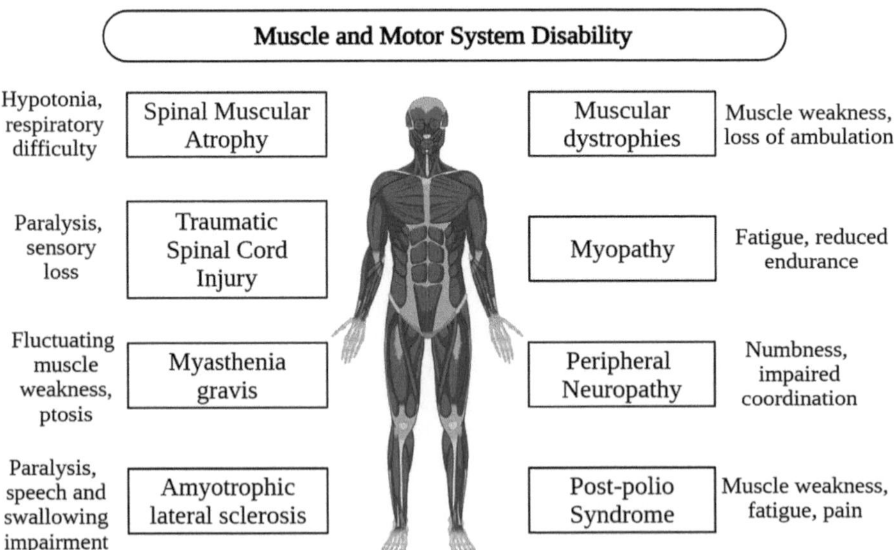

Fig. 4.1 Muscle and motor system disability

Table 4.1 Muscle and motor system disability

Disorder	Primary system affected	Underlying pathophysiology	Key functional disabilities
Muscular dystrophies	Skeletal muscle	Genetic mutations causing progressive muscle fibre degeneration	Progressive muscle weakness, loss of ambulation, and respiratory involvement
Myopathy	Skeletal muscle	Structural or metabolic muscle dysfunction	Muscle weakness, fatigue, reduced endurance
Peripheral neuropathy	Peripheral nerves	Damage to sensory and/or motor nerves	Muscle weakness, numbness, impaired coordination
Post-polio syndrome	Motor neurons	Late degeneration of neurons previously affected by poliovirus	New muscle weakness, fatigue, pain
Spinal muscular atrophy (SMA)	Anterior horn cells (spinal cord)	SMN1 gene mutation causes motor neuron degeneration	Severe muscle weakness, hypotonia, and respiratory difficulty
Traumatic spinal cord injury	Spinal cord	Mechanical damage leading to the interruption of neural pathways	Paralysis, sensory loss, and autonomic dysfunction
Myasthenia gravis	Neuromuscular junction	Autoimmune destruction of acetylcholine receptors	Fluctuating muscle weakness, ptosis, fatigue
Amyotrophic lateral sclerosis (ALS)	Upper and lower motor neurons	Progressive motor neuron degeneration	Progressive paralysis, speech and swallowing impairment

they may occur at any age. Although certain cases are inherited, the majority of adult motor system disability do not have a known cause, though some factors in the environment and infectious diseases may lead to the disease (Zambon et al., 2023).

4.2 Symptoms of Motor System Disability

Movement disorders or motor system disabilities show different signs depending on the classification. The most common one of these diseases is ataxia, which interrupts coordinated movement, making it clumsy, and causes loss of balance and speech abnormalities, possibly as a result of inherited or degenerative diseases, infections, or other controlled influences (Rosenthal, 2022). Chorea is characterized by rapid and involuntary movements of the body, primarily affecting the face and extremities; one of the major genetic causes of chorea is Huntington disease, which leads to involuntary movements, impaired thinking, and compromised psychosocial functioning (Saldert et al., 2021). Dystonia is characterized by the contraction of muscles resulting in inappropriate postures; cervical dystonia is a relatively common disorder in adults, and the muscles that are affected are those of the neck (usually the neck muscles cause the head to tilt or shake). Functional movement disorder is similar to other movement disorders, but does not occur as a result of neurological issues and responds to treatment. The multisystem atrophy is a progressive and unusual disorder that impacts several systems in the brain, resulting in conditions like ataxia or Parkinsonism, as well as such symptoms as hypotension and bladder dysfunction (Stankovic et al., 2023). Myoclonus is characterized by the rapid contraction of muscles. Parkinson's disease, being a disease of tremors, rigidity, and bradykinesia, may have non-motor symptoms like anosmia and cognitive impairment, which are slow to develop and gradually deteriorate with time. Parkinsonism, characterized by bradykinesia and rigidity, has various causes, the most common being Parkinson's disease and certain medications (Bloem et al., 2021). Progressive supranuclear palsy is a rare neurological disease that impairs gait and balance, and it can be similar to Parkinson's disease. Restless legs syndrome is a disorder that makes legs experience uncomfortable feelings as one sits, and this feeling is relieved by moving. Tardive dyskinesia comes from prolonged use of certain drugs, and this condition leads to involuntary and repetitive movements such as grimacing, blinking, etc. Tourette syndrome is exhibited by involuntary tics that appear in childhood or adolescence, whereas tremors are exhibited by rhythmic shaking, which is mainly experienced in essential tremor.

4.3 Causes of Motor System Disability

The field of genetics has a great impact on the occurrence of movement disorders, and certain disorders are caused by inherited defective DNA in the blood of parents, like Huntington disease and Wilson's disease (Wilson et al., 2025). Movement disorders may be caused by various medications, especially anti-seizure and antipsychotic medications, as well as the use of illegal drugs such as cocaine and alcohol in excess, which may lead to chorea. Ataxia can be caused by nutritional deficiencies, in particular, the deficiency of vitamins B-1, B-12, and E. There is a wide range of medical conditions, such as thyroid dysfunctions, multiple sclerosis, cerebrovascular accidents, viral encephalitis, and brain neoplasms, which can lead to movement impairments (Kramarz et al., 2024). In addition, injuries on the head can worsen these ailments. Healthcare practitioners often struggle to identify a specific cause of movement disorders, resulting in the classification of these conditions as idiopathic. The risk of developing movement disorders is significantly higher for individuals who have a family history of these disorders. Movement diseases, including essential tremor, Huntington's disease, Wilson's disease, and Tourette's syndrome, are genetic (Lange et al., 2022). Other risk factors that increase the risk of developing movement disorders include special medical-related factors, pharmacological therapies, excessive alcohol, illicit drug use (including cocaine), and vitamin deficiency. All these features increase the risk of movement problems.

4.4 Diagnosing and Treatment of Motor System Disability

In most instances, there is no available reliable testing in the case of motor system disability because the symptoms may differ in people and may even display similarity with other conditions, thus making the diagnostic process difficult. Certain disorders, such as spinal muscular atrophy (SMA), Kennedy, and certain genetic forms of amyotrophic lateral sclerosis (ALS) are tested genetically (Keinath et al., 2021). The complete diagnosis of the disease usually includes an extensive medical history, physical examination, and a more detailed neurological assessment to make measurements of motor and sensory functionalities, hearing, speech, sight, coordination, balance, and cognitive and emotional alterations. The two main tests used to diagnose possible nerve or muscle disorders, including the nerve diseases known as motor system disability, are (a) electromyography (EMG), the electrical activity of the muscles during movement and rest, which is used to analyse lower motor system disability, muscle, and peripheral nerve diseases; and (b) nerve conduction examination, a test that is usually performed with the assistance of EMG, which measures the speed and amplitude of nerve impulses through small electrodes placed on the skin (Rana et al., 2025). Lab tests, including blood, urine, or cerebrospinal fluid tests, play an important role in the detection of diseases that have related symptoms to motor system disability (Krause et al., 2021). The MRI scans are often normal in motor system disability, but are essential in the screening of other diseases like brain tumours, inflammation, infections, and vascular complications, which may lead to stroke. MRI allows the evaluation of inflammatory diseases such as multiple sclerosis, as well as brain damage. Further, magnetic resonance spectroscopy is able to assess the upper motor neuron status. Even though muscle or nerve samples may help in diagnosing the related illnesses, they are invasive, and many physicians believe that these procedures are unnecessary when diagnosing motor system disability (Christidi et al., 2022).

Currently, most motor system disability are untreatable; despite that, novel treatments are being developed (Zakharova, 2021). The prognosis depends on the course of motor system disability and the age at which symptoms appear; some forms of this disease, such as primary lateral sclerosis (PLS) and Kennedy disease, have a gradual progression and are not usually fatal, but other ones, including amyotrophic lateral sclerosis (ALS) and some forms of spinal muscular atrophy (SMA), are fatal. The primary focus of the treatment is typically on controlling symptoms, but supportive care plays a crucial role in ensuring autonomy and quality of life. Multidisciplinary clinics will be necessary in the delivery of comprehensive care of people with motor system disability. A number of drugs have been approved to treat motor system disability. Some treatments prevent the development of ALS and those that are aimed at genetic changes associated with spinal muscular atrophy (SMA) and ALS genetic variants (Wong et al., 2023). Muscle relaxants can be administered to patients as a symptom of rigidity and spasms, as well as botulinum toxin injections given to alleviate stiffness and salivation. The condition can be treated using drugs such as amitriptyline, glycopyrrolate, and atropine (Yellepeddi et al., 2024). The patients with motor system disability could use supportive therapy and assistive technology to cope with the symptoms and maintain their strength and abilities (Connolly et al., 2024). Physical and occupational therapy can improve posture, avoid joint immobility, and reduce muscle weaknesses. Stiffness may be relieved by stretching and strengthening exercises that can also enhance the circulation. It may be speech therapy that will help to overcome the challenges related to articulation, mastication, and swallowing (Pereira et al., 2024). Other support resources are the provision of heating pads to alleviate muscle pain, mobility aids, and communication devices to enhance autonomy. Proper nutrition is essential to control weight and physical power; some patients might need a feeding tube or respiratory support in their later stages in the form of a ventilator.

4.5 Different Types of Muscle and Motor System Disability

4.5.1 Muscular Dystrophies

Muscular dystrophies (MDs) are genetically inherited degenerative diseases, which are characterized by progressive muscle weakness and dystrophic alterations in muscle biopsies (Emery, 2002). The molecular genetic tests have revealed more than 50 genes related to muscular dystrophies (MDs) with categories such as congenital muscular dystrophies (CMDs) and limb girdle muscular dystrophies (LGMDs), depending on the age of onset (Table 4.2) (Kaplan & Hamroun, 2015). Conditions such as Duchenne muscular dystrophy (DMD), Becker muscular dystrophy (BMD), Emery-Dreifuss muscular dystrophy (EDMD), and facioscapulohumeral muscular dystrophy (FSHD) are well-established due to their distinctive clinical features or prevalence. It classifies major categories of illnesses and causal genes of these illnesses, where proteins are divided into extracellular matrix (ECM), sarcolemma-associated proteins, enzymatic proteins, and other categories. CMDs are mostly linked to mutations in proteins associated with the extracellular matrix and membrane proteins, while mutations in nuclear proteins lead to EDMD, and defects in sarcolemmal and sarcomeric proteins are mainly responsible for LGMDs. Combinations of muscular dystrophies and specific congenital myopathies are recommended to enhance the diagnostic processes by adding clinical presentation, age of onset, localization of muscle weakness, cardiac, laboratory (increased serum creatine kinase levels), muscle biopsy, and imaging. Followings are various proteins and their role in muscular dystrophies.

- **Simple proteins:** Deficiencies in proteins such as collagen VI, laminin 211, and alpha-dystroglycan are associated with the common types of congenital muscular dystrophies (CMDs) (Urciuolo et al., 2013). Collagen VI deficiency arises from autosomal mutations in COL6A1–3, resulting in a spectrum of severity from Ullrich congenital muscular dystrophy to milder forms like Bethlem myopathy (Demir et al., 2002). Laminin α2, produced by the LAMA2 gene, is crucial

Table 4.2 Types of muscular dystrophies

Type of muscular dystrophy	Genetic/molecular basis	Age of onset	Primary muscle groups affected	Key clinical features
Congenital muscular dystrophies (CMDs)	Mutations in genes encoding extracellular matrix or membrane proteins (e.g. laminin-α2)	At birth or infancy	Generalized skeletal muscles	Severe hypotonia, delayed motor milestones, contractures
Dystroglycanopathies	Defective glycosylation of α-dystroglycan	Congenital or childhood	Skeletal muscle, brain, eyes	Muscle weakness, intellectual disability, brain malformations
Duchenne muscular dystrophy (DMD)	X-linked mutation in dystrophin gene (DMD)	Early childhood (2–5 years)	Proximal muscles, pelvic girdle	Gowers' sign, progressive weakness, cardiomyopathy
Limb-girdle muscular dystrophies (LGMD)	Autosomal dominant or recessive mutations (various genes)	Childhood to adulthood	Shoulder and pelvic girdle muscles	Proximal muscle weakness, difficulty walking or lifting
Emery–Dreifuss muscular dystrophy (EDMD)	Mutations in emerin or Lamin A/C genes	Childhood or adolescence	Shoulder, upper arms, lower legs	Early contractures, muscle weakness, cardiac conduction defects
Facioscapulohumeral muscular dystrophy (FSHD)	Deletion or epigenetic alteration at 4q35	Adolescence or early adulthood	Facial, shoulder, upper arm muscles	Facial weakness, scapular winging, asymmetric muscle involvement

for the formation of the laminin 211 complex in muscle basement membranes. Recessive LAMA2 mutations often lead to complete protein loss, causing severe congenital symptoms, with occasional milder LGMD-like variants (Kubota et al., 2018).

- **Sarcolemma-linked proteins:** The dystrophin-associated glycoprotein complex (DAGC), which includes dystrophin, sarcoglycans, and dystroglycan, is related to Duchenne muscular dystrophy (DMD) and Becker muscular dystrophy (BMD), caused by mutations in the X-linked dystrophin gene. DAGC is linked to four autosomal recessive limb-girdle muscular dystrophies termed sarcoglycanopathies, resulting from mutations in sarcoglycan genes (Sharma et al., 2010). DAGC stabilizes muscle fibres during contraction by linking the cytoskeleton to the extracellular matrix (Bolduc et al., 2010). Dysferlin and anoctamin 5 are sarcolemmal proteins that promote the repair of the sarcolemma. Dysferlin interacts with caveolin 3, which is relevant to a rare form of limb-girdle muscular dystrophy (LGMD), while mutations in the PTRF gene lead to a deficiency of caveolin 3 and related LGMD (de Haan, 2010).

- **Enzymes:** Glycosylation abnormalities of alpha-dystroglycan (ADG) have been identified in various subtypes of congenital muscular dystrophy (CMD) and limb-girdle muscular dystrophies (LGMDs), categorized as dystroglycanopathies (Muntoni et al., 2002). ADG undergoes substantial post-translational modifications, crucial for its interaction with extracellular matrix components, including laminin (Taniguchi-Ikeda et al., 2016). Mutations affecting 18 enzymatic proteins linked to ADG glycosylation led to dystroglycanopathies, which display a range of clinical manifestations from LGMD to CMD with brain structural involvement (Godfrey et al., 2011). This highlights the importance of ADG glycosylation in muscle and cognitive function. Moreover, the role of calpain 3, a protease not associated with ADG glycosylation, in the etiology of LGMDA remains unknown (Jimenez-Mallebrera et al., 2009).
- **Nuclear membrane proteins:** The nuclear envelope comprises two membranes that engage with the nuclear lamina. Mutations in nuclear envelope proteins, including lamin A/C, emerin, nesprin 1 and 2, and TMEM43, are linked to many types of muscular dystrophy, including Emery-Dreifuss muscular dystrophy (EDMD) (Bonne et al., 2000). Additionally, mutations in matrin 3 are associated with distal myopathy and vocal cord paralysis (Foley et al., 2011). Phenotypic changes arising from these mutations indicate distinct functions of protein domains, with lamin A/C mutations linked to various illnesses such as EDMD, isolated cardiomyopathy, limb-girdle muscular dystrophy with lipodystrophy, and progeria, among others (Bertrand et al., 2011).
- **Sarcomeric proteins:** Recent results indicate that several non-dystrophic congenital myopathies contribute to dystrophic muscle disorders, particularly those associated with the titin gene (TTN). The overall knowledge of its phenotypes is largely advanced due to next-generation sequencing, thanks to its sheer scale (Jablonka et al., 2022). Defects in sarcomeric proteins often cause significant distal weakness, which is different from most diseases related to congenital muscular dystrophies (CMDs) and limb-girdle muscular dystrophies (LGMDs) (Yoskovitz et al., 2012). In certain cases, especially in TTN and MYH7, severe dilated cardiomyopathy could present with muscle symptoms; however, alterations in these genes could also cause the occurrence of isolated cardiomyopathy.
- **Endoplasmic reticulum proteins:** TRAPFC, which is an element of the transport particle protein complex (TRAPMC), has been associated with limb-girdle muscular dystrophy (LGMD2S) and congenital muscular dystrophy (CMD), with one that includes cerebellar involvement (Bögershausen et al., 2013). Moreover, defects of the BVES protein that locates itself in the endoplasmic reticulum and sarcolemma result in an autosomal recessive form of limb-girdle muscular dystrophy (LGMD), which causes progressive weakening and heart fibrillation disorders (Schindler et al., 2016).
- **Other proteins:** FSHD1 is a common autosomal dominant muscular dystrophy, which is associated with loss of major copies of the D4Z4 repeat in the 4q35 subtelomeric region, causing inadequate repression of the transcription factor DUX4 (Lemmers et al., 2010). This loss, in combination with the presence of a

particular polymorphic form providing the DUX4 transcripts with a canonical polyadenylation signal, is required to cause cellular toxicity (Wallace et al., 2012). Conversely, FSHD2 is due to de-repression of DUX4 that occurs through hypomethylation of the same site on the chromosome as a result of mutations in the SMCHD1 gene located on chromosome 18p11. This inconsistency shows the genetic complexity that lies in the nature of both forms of FSHD and depends on changes that occur in the 4q35 region.

4.5.1.1 Congenital Muscular Dystrophies

This term refers to children who exhibit signs of LAMA2-CMD at birth or during the first year and have overall hypotonia, limited mobility, and elevated blood CK concentrations. Early clinical symptoms include upper limb and axial weakness, followed by lower limb weakness and contractures. Affected children can sit on their own but rarely walk or stand; additionally, they have problems with swallowing and breathing complications to the point that they need to be ventilated by adolescence. Scoliosis is widespread, but there is no significant impairment of cognitive functioning in spite of white matter that is observed on MRI. Focal epilepsy is present in about 20% of the cases, and the quality of care has a significant impact on the prognosis, as a large number of patients survive to early adulthood. The diagnosis depends on significantly decreased muscle or skin protein expression and has to be confirmed genetically (Rajput et al., 2023). Ullrich COL6-related congenital muscular dystrophy (CMD) may present at birth, characterized by skeletal deformities such as kyphoscoliosis, pronounced joint contractures, limb hyperlaxity, torticollis, and hip dysplasia. The creatine kinase levels are often in the normal range or slightly high. About half of the patients learn independent ambulation, but this ability tends to decrease in adolescence because of the gradual muscular weakness and the rigidity of joints (Yonekawa & Nishino, 2015). Gastrostomy due to failure to thrive and nocturnal respiratory support are frequently required because respiratory insufficiency occurs during early adolescence. There are no cognitive and cardiac alterations. Reduced collagen VI expression in muscle fibres or fibroblasts, together with severe dystrophic alterations and characteristic muscle MRI appearances, confirms that collagen VI is involved in the diagnosis. The genetic study of COL6A1, COL6A2, and COL6A3 identifies recessive or de novo dominant mutations, with a recent finding of a de novo dominant mutation in COL6A1 as one of the most common in this disorder (Bolduc et al., 2019).

4.5.1.2 Dystroglycanopathies

Dystroglycanopathies are a diverse group of diseases that have different severity in the case of recessive mutations in glycosylation of dystroglycan. Reduced glycosylation primarily affects skeletal or cardiac muscle, resulting in more serious defects that result in problems in the central nervous system, such as cognitive impairment and significant cortical dysplasia. Severe disorders such as muscle-eye-brain disease, Fukuyama congenital muscular dystrophy, and Walker-Warburg syndrome are structural ocular defects, and they are usually associated with seizures and visual impairment. Diagnosis is usually made with high levels of CK, dystrophic muscle biopsies, and a low concentration of glycosylated alpha dystroglycan at the sarcolemma, which is backed up by genetic analysis of the relevant genes.

4.5.1.3 Duchenne Muscular Dystrophy

Duchenne Muscular Dystrophy (DMD) is commonly found in male sufferers at early stages of infancy and is marked by such features as frequent falls and difficulties with running or climbing stairs, as well as the Gowers manoeuvre used to stand up. About half exhibit delayed speech, and about one in five show comorbidities in the form of intellectual disability, autism, and attention deficit problems (Ricotti et al., 2016). The condition can remain stable until around age seven, after which it rapidly progresses, leading to loss of ambulation by age 12; it is often comorbid with scoliosis and cardiomyopathy, and individuals typically have an average survival into their late teens. Advances in respiratory and cardiac interventions have now enabled walking for individuals between the ages of 13 and 14, and life expectancy has been improved to reach the late twenties of their lives. Progressive dilated cardiomyopathy is now leading to higher death rates (Cheeran et al., 2017). Recent studies suggest that eplerenone may benefit overall heart health. However, it remains uncertain whether its preventive use would be effective for patients under the age of 10, necessitating further intensive research (Raman et al., 2015). The involvement of the brain in muscular dystrophies, unlike the progressive degeneration depicted in the skeletal and heart muscles, is stable. Muscle tissues produce a single isoform, whereas the brain contains numerous dystrophin isoforms. The difference in the involvement of the cerebral side is because the mutation site of the DMD gene is unique, and thus it affects the amount of brain dystrophin transcripts (Battini et al., 2018). The emergence of clinical trials has brought out the importance of validated outcome measures and intensive natural history explorations. The major outcome measure of many studies is the six-minute walking test (6MWD); disease-specific functional assessments, including the North Star Ambulatory Assessment and the Performance of Upper Limb Test, are gaining popularity. The studies of natural history show that the course of Duchenne Muscular Dystrophy (DMD) is non-linear and has a high degree of variability even between years (Goemans et al., 2013). Functional tests can be complemented by imaging to better distinguish between patients at different levels of illness, e.g. MRI. The classifications of the DMD progression trajectories have been improved by the recent statistical methodologies that establish the role of various factors, including age and timed function assessments, in defining the trajectories (Mercuri et al., 2016).

4.5.1.4 Limb Girdle Muscular Dystrophies

The heterogeneity of limb-girdle muscular dystrophies (LGMDs) in both clinical and genetic aspects and overlapping features with other disorders makes the process of diagnosis more complicated. The variants usually appear after ambulation has started, with the proximal muscle weakness sparing the face and additional extra-ocular muscles and dystrophic changes that are observable in muscle biopsies (Murphy & Straub, 2015). Diagnostic methods include the observation of certain proteins in muscle biopsies and the images observed in muscle imaging. There are great differences within the subtypes, especially when respiratory and cardiac are involved, and as such, genetic identification is needed (Savarese et al., 2015). The next-generation sequencing process is becoming a dominant research tool, and a recent workshop recommended a new nomenclature for LGMDs (Straub et al., 2018).

4.5.1.5 Emery Dreifuss Muscular Dystrophy

The phenotypical features of Emery-Dreifuss muscular dystrophy (EDMD) are progressive muscle weakness and atrophy, common to the biceps and calves, and frequent contractures in the elbows, Achilles's tendons, neck, and spine. It usually begins at the early adolescence stage, and it is accompanied by problems like atrial and cardiac conduction abnormalities, ventricular arrhythmias, and dilated cardiomyopathy that affect long-term survival. Concerning the defibrillator implantation, the recommendations of care are that since pacemakers have limited impact with regard to survival, their implantation is advised. In contrast to X-linked emerin-mutation EDMD, dominantly acting LMNA mutation-related EDMD presents itself earlier; some children with the condition display signs of congenital muscular dystrophy. Such children might never be able to walk, or they might lose this control when they enter their first decade, with a significant loss of muscularity and with the development of respiratory and heart problems. The levels of serum CK tend to be elevated, and the muscle MRI indicates a characteristic distribution of the selective involvement. Clinical diagnosis can be essential by means of genetic verification, in which the relevant genes are examined.

4.5.1.6 Facio Scapulo Humeral Muscular Dystrophy

Facioscapulohumeral Muscular Dystrophy (FSHD) is characterized by progressive and often asymmetrical weakness of the face, scapular, proximal limb, and peroneal muscles, which is usually associated with the drop of the foot. There are common features such as scapular winging with facial weakness, but cardiorespiratory function is usually normal (Goselink et al., 2016). The weakness is not regular; most people can walk as adults, while others lose much of their weakness and mobility in their teens, especially if symptoms began in childhood. Financial vasculopathy and sensorineural hearing impairment are also features of FSHD. The level of creatine kinase (CK) is usually within the normal range or slightly high, and genetic diagnostics is directed at the chromosome 4q35 or the FSHD2 locus (Mercuri et al., 2019).

4.5.2 Myopathy

Myopathies represent an extensive range of diseases, which mainly affect skeletal muscle structure, metabolism, or channel functioning and result in muscular weakness that impedes daily activities. Symptoms such as muscular weakness, stiffness, cramps, and spasms are common, inclusive of the soreness of muscles. Some myopathies can be associated with rhabdomyolysis. The myopathies are typified by motor dysfunction and typically in the form of proximal muscular weakness in the area of the pelvic and shoulder girdle, with the muscles of the pelvis usually being more severely affected. Diagnosis requires a comprehensive history and a physical assessment. The patients might have difficulty in standing up when in a seated position, climbing stairs, or making overhead movements. Other muscular parts, such as the thighs, back, or fingers, can be involved with or without such symptoms as myalgia, rashes, fatigue, or cramps. In extreme situations, rhabdomyolysis can be experienced, which results in dark urine following excessive physical activity.

Autoimmune disorders such as polymyositis and dermatomyositis are more common in women and are characterized by proximal muscle weakness, primarily affecting the pelvic girdle rather than the shoulder girdle. Polymyositis is combined with arthralgia, whereas dermatomyositis has other manifestations; in addition to skin changes, heliotrope rash (purple rash on the eyelids), Gottron papules (erythematous scaly rash on the fingers), and Shawl sign (reddish rash on the shoulders and the back) can be observed. Dermatomyositis can also include interstitial lung disease, gastrointestinal vasculitis, and paraneoplastic syndrome augmented to underlying cancer (Ofori et al., 2017). Hypothyroid and hyperthyroid myopathies are also associated with thyroid disorder, which is marked by proximal muscle weakness and peripheral neuropathy. Hypothyroid myopathy has such characteristics as pseudohypertrophy, myoedema, and tardy deep tendon reflexes. In contrast, hyperthyroid myopathy is linked to Grave's ophthalmopathy, goiter, and weakening of the extraocular muscles (Szczęsny et al., 2018; Vignesh et al., 2013). Other acquired myopathies, such as sarcoidosis, amyloidosis, and critical illness myopathy, exhibit unique clinical symptoms. Inherited myopathies, including Duchenne muscular dystrophy, myotonic dystrophy types 1 and 2, mitochondrial myopathies, and glycogen storage diseases such as McArdle and Pompe's disease, are rare in children and often lead to considerable systemic complications, consequently restricting life expectancy based on disease severity and related complications (Cardamone et al., 2008).

4.5.2.1 Treatment and Management

The care of myopathy mostly involves supportive measures, focusing on exercise, physical and occupational therapies, dietary support, and, at times, genetic counselling (Voet et al., 2019). In instances of hereditary muscular dystrophies, prednisone (0.75 mg/kg/day) may enhance muscle strength and slow illness development. Mitochondrial myopathies may exhibit a favourable response to creatine monohydrate (5–10 g/day), although Coenzyme Q10 need further research (Voet et al., 2019). Acquired myopathies improve with the treatment of underlying conditions, such as thyroid abnormalities or infections. Management of toxin-induced myopathy involves the removal of the causative agent. HIV-associated myopathy responds to antiretroviral therapy and maybe corticosteroids (Katzberg & Kassardjian, 2016). Inflammatory and autoimmune myopathies are treated with immunomodulatory and immunosuppressive medications, including methotrexate, azathioprine, and corticosteroids. It is reported that the cases are resistant, and tough clinical trials are required. The main priorities in managing rhabdomyolysis are to prevent acute kidney damage through intensive hydration and to maintain renal function and electrolyte balance.

4.5.2.2 Differential Diagnosis

Guillain-Barre syndrome (GBS) is a neurological disorder that manifests 2–4 weeks following a minor respiratory or gastrointestinal infection, whose symptoms are characterized by the presence of finger dysesthesias and muscle weakness of the lower extremities (proximal). This weakness is fast to advance, affecting the arms, torso, cranial nerves, and respiratory muscles (Nagy & Veerapaneni, 2023). The diagnosis of tick-borne diseases is based on the indirect immunofluorescence antibody (IFA) technique on paired serum samples taken at the beginning and 2–4 weeks after. The Lambert-Eaton myasthenic syndrome (LEMS) is a condition of proximal muscular weakness, reduced tendon reflexes, post-tetanic potentiation, and autonomic alterations caused by non-release of acetylcholine. Myasthenia gravis is an autoimmune disease that affects peripheral nerves, which is caused by antibodies that target acetylcholine (ACh) nicotinic receptors in the neuromuscular junction. This leads to a reduction in ACh receptors, resulting in progressive muscle weakness with repeated use, which is partially restored through rest. Bulbar muscles are the most affected muscles; however, patients tend to have varying amounts of general weakness.

4.5.3 Peripheral Neuropathy

Peripheral neuropathy is a disease affecting about 1% of the adult population in the global population and is characterized by damage to the peripheral nerves, symptoms of which include moderate levels of numbness of the toes and in severe cases, may need wheelchair support. The incidence is higher in old age, ranging between 6% and 10% among patients aged over 60 (Hanewinckel et al., 2016; Hoffman et al., 2015). The seventh most common neurological disability is diabetic neuropathy, which affects about 206 million people. The causes are most frequently diabetes, alcohol use, and neurotoxic treatment in developed nations and infectious diseases like leprosy and HIV in low-income countries. The most prevalent type of this disease identified by general practitioners is length-dependent peripheral neuropathy (LDPN), which is mostly sensory and progresses from the toes to more proximate areas, including the knees and the hands (Rodrigues & Lockwood, 2011). Length-dependent polyneuropathy (LDPN) normally starts with peripheral nerve damage, and it manifests as numbness and tingling in the toes and feet, which slowly spreads symmetrically down the lower limbs over weeks to months. There is a possibility of neuropathic pain, and with the disease coming up, pain might extend to the knees and even the fingertips. Mild muscle weakness can manifest in toes, and autonomic signs, particularly when a patient has diabetes or amyloidosis, may include the absence of sweating (anhidrosis), disorientation, and gastrointestinal issues (Loavenbruck et al., 2016). Clinicians are obliged to examine such common underlying conditions as diabetes, alcohol use disorder, and vitamin B12 deficiency, but they need to be aware of drugs that negatively influence nerve functioning (Winstock & Ferris, 2020). A detailed assessment of the presence of foot abnormalities and foot neuropathy history is necessary, as a significant percentage of idiopathic neuropathies are hereditary and often in connection with Charcot-Marie-Tooth disease (Dyck et al., 1981).

Clinicians should also examine the sensation at the great toe and index finger and progressively proceed toward the proximal position until the normal sensation occurs. The assessment includes big fibre functions (vibration, proprioception, light touch) and small fibre functions (pinprick, temperature). A positive Romberg sign is an indicator of proprioceptive dysfunction. The motor assessment can also show foot atrophy and toe weakness, which affects the ankles and hands. Persons with LDPN often have ankle-deep tendon reflexes of reduced or absent strength. In mild sensory-predominant LDPN, gait is generally normal, but in severe cases, the gait may have an unsteady, wide-based gait that is due to sensory ataxia. Charcot-Marie-Tooth Dental pathology is often characterized by weakness of the distal foot and ankle, absence of neuropathic sensory symptoms, and morphological defects like pes cavus and hammertoes. Atypical features of LDPN include pronounced motor weakness, significant autonomic symptoms (orthostatic intolerance, postprandial nausea, vomiting, diarrhoea), symptom onset and progression over days to weeks, non-length-dependent sensory loss or weakness, asymmetrical neuropathy, and systemic manifestations (weight loss, fever, rash) (Watson & Dyck, 2015). These signs could be attributed to the presence of an underlying immunological, inflammatory, infectious, or neoplastic disease and require urgent consultation with specialists, such as neurologists, rheumatologists, specialists in infectious diseases, or cancer specialists. The missed atypical presentations may lead to delayed treatment and poor patient outcomes. Table 4.3 illustrates various types and their cause of peripheral neuropathy.

Table 4.3 Types and causes of peripheral neuropathy

Type/cause	Underlying aetiology	Commonly affected nerves	Key clinical features
Peripheral neuropathy (general)	Damage to peripheral nerves due to metabolic, nutritional, genetic, or immune causes	Sensory, motor, and autonomic nerves	Numbness, tingling, weakness, pain
Neuropathy associated with diabetes	Chronic hyperglycaemia causes microvascular and nerve damage	Sensory and autonomic nerves	Stocking–glove sensory loss, neuropathic pain
Neuropathy due to nutritional imbalances	Deficiency or excess of essential nutrients	Sensory and motor nerves	Paraesthesia, weakness, gait disturbances
Vitamin B12 deficiency neuropathy	Impaired myelin synthesis	Sensory and motor nerves	Loss of vibration sense, ataxia, and cognitive changes
Copper deficiency neuropathy	Impaired mitochondrial and myelin function	Sensory pathways	Gait instability, sensory loss
Vitamin B6 (pyridoxine) neuropathy	Deficiency or toxicity affecting nerve metabolism	Sensory nerves	Sensory ataxia, numbness, paraesthesia
Thiamine (vitamin B1) deficiency neuropathy	Impaired energy metabolism in neurons	Motor and sensory nerves	Weakness, burning feet, peripheral pain
Hereditary neuropathies	Genetic mutations (e.g. Charcot–Marie–tooth disease)	Motor and sensory nerves	Distal muscle wasting, foot deformities
Autoimmune and inflammatory neuropathies	Immune-mediated nerve damage (e.g. Guillain–Barré syndrome)	Motor and sensory nerves	Rapid weakness, areflexia, sensory loss

4.5.3.1 Neuropathy Associated with Diabetes

Diabetes is the major cause of global peripheral neuropathy cases, with 20 to 40% of diabetic patients exhibiting indications of diabetic peripheral neuropathy (LDPN) (Dyck et al., 1993). The objective assessment data indicate that among the patients with diabetes, about 50% have peripheral neuropathy (Liu et al., 2019). The risk factors for LDPN are long-term diabetes, old age, high levels of HbA1c, diabetic retinopathy, obesity, and hyperlipidaemia (Cortez et al., 2014). Recent-onset diabetes is usually characterized by disagreeable foot-related sensory problems and metabolic syndrome. Compressive mononeuropathies are common causes of such problems as carpal tunnel syndrome in diabetic individuals (Sanjari et al., 2024). LDPN is caused by chronic hyperglycaemia and is an illness that causes sensory, motor, and autonomic impairment. Treatment must be effective, implying that there should be control of glucose levels as well as control of obesity, hyperlipidaemia, and hypertension (Elsayed et al., 2024). Glycaemic control is more effective in preventing peripheral neuropathy in type 1 diabetes than in type 2 diabetes, where its potential effect on symptoms and progression is limited

(Callaghan et al., 2012). A type of diabetic LDPN, insulin neuritis is a diabetic neuropathy resulting from treatment and can be associated with a significant improvement in HbA1c (above 2% in 3 months) (Malviya & Rajput, 2025). It is being marked by intense and debilitating pain due to small fibre neuropathy and autonomic neuropathy, which appear within 8 weeks of such re-reduction. A 2015 retrospective study found that 10.9% of patients with diabetes have this condition, and an increase in HbA1c greater than 4 percentage points in 3 months is related to the fact that an increase is more than 80% likely to develop neuropathy (Gibbons & Freeman, 2015). The inflammatory diabetic radiculoplexus neuropathy, an uncommon and challenging condition associated with type 2 diabetes, affects the lumbar plexus as an abrupt unilateral painful neuropathy, leading to significant muscle weakening and atrophy (Dyck et al., 1999). In a 2019 study, the frequency of annuity in Olmsted County, Minnesota, was evaluated to be 2.79 per 100,000 inhabitants yearly (Ng et al., 2019).

4.5.3.2 Neuropathy due to Nutritional Imbalances

- **Vitamin B12:** Vitamin B12 deficiency is a rare disorder that occurs in people who avoid animal food and also in people with issues in the absorption of vitamin B12, like atrophic gastritis, or after gastric bypass surgery. The symptoms can be a loss of sensation, paraesthesia, and ataxia, along with central nervous system damage-induced cognitive deficits. It may include treatment of 1 mg of vitamin B12 by subcutaneous or intramuscular injection each week during the first month, or monthly, but higher doses of oral supplements (1–2 mg/day) are as effective (Wang et al., 2018).
- **Copper:** Copper deficiency can be mistaken for B12 deficiency symptoms, especially in patients experiencing sensory ataxia for extended periods, ranging from weeks to months. It is associated with malabsorption of the small intestine or excessive intake of zinc that could be treated with oral or parenteral supplementation of 2–4 mg per day (Taylor et al., 2017; Kumar et al., 2004).
- **Vitamin B6 (Pyridoxine):** High levels of vitamin B6 can manifest as permanent sensory neuropathy, particularly of the dorsal root ganglia, resulting in sensory ataxia. Toxicity is usually linked to excessive intake that is mostly through supplements, exceeding the intake of 2 g daily; however, neuropathy has been observed with continued intake of 50 mg per day (Berger et al., 1992). Deficiency in vitamin B6, which is often caused by certain medications, dialysis, autoimmune conditions, or malabsorption, can be corrected through supplementation, but the evidence to prove its ability to prevent peripheral neuropathy in that situation is lacking. The suggested supplementation dosage will be 50 mg per day, which is only applicable in specific conditions of deficiency or prevention (Muhamad et al., 2023).
- **Thiamine:** Thiamine (vitamin B1) deficiency causes wet (cardiac failure) or dry (neuropathic) beriberi, which shows similar symptoms. Despite the limited epidemiological evidence, beriberi is uncommon in high-income countries, with the

majority of people affected having moderate alcohol use disorder or malnutrition and, in particular, after bariatric surgery (Koike et al., 2008). It can be axonal LDPN or polyradiculoneuropathy, like in Guillain-Barré syndrome, and can also be Wernicke-Korsakoff syndrome, characterized by confusion, ataxia, ocular movement difficulties, memory impairment, and confabulation. The management involves the use of thiamine as a prelude to glucose to prevent the worsening of the deficiency, especially in Wernicke-Korsakoff syndrome, using acute replacement doses of 100 to 200 mg intravenously three times daily over 3 days, followed by 100 mg orally as long-term maintenance. People who are at greater risk are to receive a 100 mg oral dose of thiamine daily.

4.5.3.3 Hereditary Neuropathies

The severity of hereditary neuropathies is of a different nature, with the Charcot-Marie-Tooth disease (CMT) being the most common type. CMT affects both motor and sensory neurons, thereby weakening and atrophying the muscles, mostly in the lower extremities, with some cases leading to abnormalities of the feet. Its prevalence is between 10 and 30 per 100,000 people and is typically determined among patients with peripheral neuropathy and those with a family history of the disease (Skre, 1974). The symptoms generally appear before the age of 20, and over 100 genetic aetiologies are known, which leads to a couple of subtypes (Murphy et al., 2012). Diagnosis entails electrodiagnostic testing to distinguish between axonal and demyelinating types, alongside genetic testing principally for PMP22 duplications or deletions, succeeded by next-generation sequencing to identify further harmful mutations (Record et al., 2024). Hereditary neuropathies may affect motor nerves (distal hereditary motor neuropathies), sensory and autonomic nerves (hereditary sensory and autonomic neuropathies), or both systems (spinocerebellar ataxias and hereditary spastic paraplegias) (Currò et al., 2021). A prevalent etiology of sensory length-dependent peripheral neuropathy (LDPN) is the pathogenic repeat expansion of the RFC1 gene. Hereditary transthyretin amyloidosis, resulting from mutations in the transthyretin gene, can lead to length-dependent sensory neuropathy and autonomic manifestations, including orthostatic hypotension and diarrhoea. Although the majority of inherited neuropathies do not have targeted treatments, hereditary transthyretin amyloidosis can be treated with stabilizing agents (diflunisal, tafamidis) and drugs that diminish transthyretin production (patisiran, vutrisiran, eplontersen) (Benson et al., 2018). A randomized investigation showed that 29.7% of patients receiving diflunisal achieved neuropathic stabilization, compared to 9.4% in the placebo group (Conte et al., 2013). Phase 3 trials of treatments aimed at reducing transthyretin production demonstrated substantial enhancements in neuropathy impairment and quality of life measures at 15 or 18 months relative to placebo (Coelho et al., 2023).

4.5.3.4 Autoimmune and Inflammatory Neuropathies

Neuropathies resulting from autoimmune or inflammatory disorders, albeit uncommon and generally not categorized as LDPN, should be recognized as they may respond to treatments like prednisone or intravenous immunoglobulin (Hattori et al., 1999; Tracy, 2023). A neuromuscular neurologist examination is essential for individuals with rapidly progressing neuropathy to ensure a precise diagnosis and initiate tailored treatment plans that could prevent progression and improve peripheral neuropathy (Mauermann & Staff, 2025).

4.5.4 Post-Polio Syndrome

Poliomyelitis was a physically incapacitating infection in the twentieth century, and more so in the 1940s and 1950s, and there was a significant epidemic in India in 1988 (Cousins, 2017). The polio vaccination led to the reduction of new infections by 99%. However, 1520 million people in the world still show the effects of the virus, especially the Post-Polio Syndrome (PPS) that can be realized after 15 years of stasis. PPS includes new muscle weakness, atrophy, fatigue, and pain, which greatly negatively affect the quality of life. Post-Polio Syndrome (PPS) occurs in a range of 20 to 85% among polio survivors, indicating that the syndrome can still develop later, even though cases of acute polio have decreased. PPS has not been sufficiently studied and understood (Farbu et al., 2006). In acute poliomyelitis, 95% of all patients are either asymptomatic or have the mild flu-like symptoms, and 5% of the patients advance to the paralytic type, which is characterized by the involvement of the spinal muscles that affect the limbs and breathing muscles. Poliovirus type 1 is the main causal agent that is able to cross the blood-brain barrier autonomously, leading to the inflammation of meninges, spinal cord, and brain (Leon-Monzon & Dalakas, 1995). The initial pathogenic feature of paralytic poliomyelitis is the destruction and death of the anterior horn followed by a prolonged period of axonal growth, which may last as long as thirty years, leading to the symptoms of post-polio syndrome (PPS) (Luciano et al., 1995). This process involves the process of expanding the motor units that can become biologically not viable. The causes of motor unit failure include metabolic stress, excessive use, age, and chronic inflammation, and they are all related to the functional decline observed in longitudinal studies (Wiechers & Hubbell, 1981).

The muscular fatigue and myalgia associated with PPS are linked to cellular changes in the muscles and alterations in contractile properties due to the transformation of type II (fast) fibres into type I (slow) fibres (Ramachandran et al., 2021). The presence of the persistent or reactivated poliovirus in survivors has been a debatable issue because a number of studies have detected the presence of viral sequences and elevation of serum IgM anti-poliovirus antibodies, although others claim no supportive evidence (Sandberg, 2023). Increased expression of pro-inflammatory cytokines, including TNF-a and IFN-g, in PPS patients post-mortem and the possibility of an inflammatory or autoimmune aspect of PPS, but no definite associations between any of these factors and symptom severity or disease course (On et al., 2024). Biopsies of skeletal muscle of PPS patients are characterized by inflammatory signals; however, there is little evidence of an autoimmune etiology (Melin et al., 2014). PPS genetic predisposition is unclear, but there are many variations that may be involved; there is no single risk profile that is conclusive. The pathophysiology of PPS is complex, and it involves a combination of inflammatory, immunological, and genetic factors (Querin et al., 2019). Contradictory results on the involvement of the brain in post-polio syndrome (PPS) are presented in postmortem studies (Li et al., 2021). The 50- to 70-year-old historical research indicates that the poliovirus attacks different parts of the brain, including the reticular formation, hypothalamus, thalamus, putamen, caudate, locus coeruleus, and substantia nigra, which can possibly explain the symptoms of fatigue and attention deficit later on in life (Russo & Riessland, 2022). However, recent studies have shown that there is no cerebral occurrence, but rather selective spinal cord disease bypassing the dorsal roots, and the manifestation of enterovirus RNA was only found in the spinal cord (Jellinger, 2025). Some cases of polio patients have also shown amyotrophic lateral sclerosis (ALS) along with specific histological changes. Further investigation and research on the brain and spinal cord in post-polio syndrome have a massive gap compared to other motor system disability. The utilization of magnetic resonance imaging (MRI) to measure these changes in volume and their correlation with those observed after the organism was found dead has primarily been required to demonstrate the occurrence of hyperintensities in specific regions of the brain that correlate with prior studies (McKenna et al., 2021). Extensive comparative research studies have shown that there was no significant deterioration in the size of the brain in PPS patients compared to multiple sclerosis patients and controls, and that there was no relationship between weariness and brain size (Li et al., 2021). The majority of modern research has been cross-sectional and thus can only give insights into longitudinal variations, but a longitudinal case-control study is currently being conducted to clarify the spinal cord alterations in PPS (Benatar et al., 2022).

4.5.4.1 Clinical Manifestations of Post-Polio Syndrome

The post-polio syndrome (PPS) is defined by the development of muscular weakness, loss of endurance, muscle atrophy, muscle pains, and fasciculations. The other symptoms are pervasive fatigue, cold intolerance, dysarthria, dysphagia, and respiratory problems, particularly in previously affected parts and subclinically affected regions (Tanrıverdi et al., 2025). Assistive equipment is sometimes required to overcome the limitations related to ambulation, and it is more prone to falls (Toniolo et al., 2025). PPS is also linked with a range of non-motor symptoms, including sensory deficit and paraesthesia, which are tied to alterations in sensory evoked potentials, which pertain to atrophy of the spinal cord (Li et al., 2021). PPS patients have cognitive impairments such as difficulties in word retrieval, decreased concentration, lack of attention, memory loss, and mood swings that are not sufficiently assessed despite their influence on the quality of life (Aljarallah et al., 2023). These characteristics result in fewer social interactions and even social isolation. One of the widespread and alarming concerns, generalized fatigue, is supposed to be the result of a complex of muscular diseases, obesity, respiratory dysfunction, polypharmacy, and sleeping problems. It is important to identify the specific causes of fatigue to manage it properly (Li et al., 2021). A range of sleep disorders, including restless leg syndrome (RLS), obstructive sleep apnea (OSA), excessive daytime somnolence (EDS), and periodic limb movements, are common in post-polio syndrome (PPS), with restless leg syndrome being strongly linked to the severity of fatigue (Ding et al., 2025). The fact that the severity of fatigue correlates with RLS, and the response to treatments, including pramipexole, also shows potential pathophysiological connections (Erokhina et al., 2024). PPS has reported an increase in the number of cases of cauda equina syndrome and renal impairment, but the relationships between them are poorly investigated (Tsai & Lin, 2025; Li Hi Shing et al., 2019).

4.5.5 Spinal Muscular Atrophy

Spinal muscular atrophy (SMA) is a group of hereditary diseases characterized by a loss of anterior horn cells, the degeneration of muscles, and their weakness. The most common form, an autosomal recessive disorder, is caused by mutations of the SMN1 gene on chromosome 5q13, which comprises more than 95% of the cases. One study has suggested a carrier rate of 1 in 54 and an incidence of 1 in 11,000 (Sugarman et al., 2012). SMA ranges in severity, with clinical features being described in four phenotypes based on the age of onset and motor ability (Table 4.4). Although it is still not possible to locate a solution, advancements in molecular genetics have enabled the creation of treatment options that are currently in the early stages of clinical trials (Lorson et al., 2010). However, in spinal muscular atrophy (SMA), understanding of its natural history, where early detection and treatment intervention are paramount, is of enormous importance, particularly in the severe type 1, which causes significant motor and respiratory degradation during the first year (Finkel et al., 2014). Non-invasive ventilatory support and enteral nutrition would be effective in enhancing survival rates of over one year to 70% or more. On the other hand, less severe forms (type 2 and type 3) show few changes in motor and respiratory functions over a period of one year (Kaufmann et al., 2011). The main clinical features of Spinal Muscular Atrophy (SMA) include muscular weakness and atrophy, which are, as a rule, symmetrical and affect mostly the proximal muscles instead of the distal ones. The pathophysiology is linked to degeneration of anterior horn cells, which causes proximal weakness of the extremities and influences axial, intercostal, and bulbar muscles. An international consortium used a taxonomy in 1991 to define three levels of SMA depending upon the level of motor functioning and the age of onset (Dunaway Young et al., 2023). The later

Table 4.4 Classification of Spinal Muscular Atrophy (SMA)

SMA type	Age of onset	Genetic basis	Motor milestones achieved	Key clinical features
SMA type 0	Prenatal or at birth	Homozygous deletion or mutation of *SMN1* gene	None	Severe hypotonia, respiratory failure at birth, joint contractures
SMA type 1 (Werdnig–Hoffmann disease)	Infancy (0–6 months)	*SMN1* gene deletion/mutation	Never able to sit independently	Profound muscle weakness, hypotonia, feeding and breathing difficulties
SMA type 2 (intermediate SMA)	6–18 months	*SMN1* mutation with partial *SMN2* compensation	Able to sit but not walk independently	Progressive muscle weakness, scoliosis, respiratory compromise
SMA type 3 (Kugelberg–Welander disease)	Childhood or adolescence	*SMN1* mutation with higher *SMN2* copy number	Able to walk initially	Proximal muscle weakness, difficulty climbing stairs, fatigue

developments defined differences in type 3 according to age of onset, type 4 of adult cases, and type 0 of prenatal onset (Nishio et al., 2023). Despite the challenges in the correct classification, this approach is essential to provide clinical and prognostic information.

4.5.5.1 SMA Type 0

Spinal muscular atrophy type 0 is characterized by severe weakness and hypotonia in newborns, which is presumably prenatally based. Affected infants have areflexia, facial diplegia, atrial septal defects, and joint contractures. Loss of life respiratory failure is a severe threat with a significant reduction of life expectancy, with most infants dying within less than 6 months (Mežnarić et al., 2025).

4.5.5.2 SMA Type 1

Werdnig-Hoffman disease, or type 1 SMA, is a disease that occurs in infants and is manifested by hypotonia, poor head control, and the absence of tendon reflexes before the age of 6 months. They cannot sit up on their own or can sit in a frog-leg posture. The injury of the intercostal muscles causes the bell-shaped thorax and paradoxical breathing. The onset of lingual and swallowing paralysis is common and is succeeded by tongue muscular tremors and eventual paralysis of the face. Such infants are also associated with aspiration and inability to thrive, with respiratory failure being experienced at a very young age of two (Cances et al., 2022). As there is attention and focus on the newborns at the diagnosis, the cognitive function is largely preserved.

4.5.5.3 SMA Type 2

Children who have type 2 SMA can sit on their own, yet cannot walk independently. The variant of SMA is characterized by the progression of proximal weakness in the legs, hypotonia, and areflexia. Other orthopaedic comorbidities, caused by muscular weakness, are scoliosis, joint contracture, and ankylosis. Scoliosis and weakness of intercostal muscles play off against each other and could lead to restrictive lung disease, but do not impair cognitive functioning (Zappa et al., 2021).

4.5.5.4 SMA Type 3

People, both children and adults, with type 3 SMA, also known as Kugelberg-Welander's disease, may be able to walk independently. However, they can experience progressive muscular weakness, particularly in the lower limbs, which may necessitate the use of a wheelchair. Unlike type 2 SMA, they do not normally show scoliosis, and they have a slight deterioration in respiratory muscle function. The life span and cognitive abilities of individuals in this group remain stable (Annoussamy et al., 2021).

4.5.5.5 Molecular Genetics

SMA posed a challenge before the clarification of its genetic basis because of the variability of severity caused by a single gene mutation. In 1995, the Melki laboratory identified a deletion of 95% of cases of spinal muscular atrophy (SMA) were caused by the deletion of a homozygous deletion in the SMN1 gene that is situated on chromosome 5q13 (Huang et al., 2024). Human beings possess two different forms of the SMN gene: telomeric SMN1 and centromeric SMN2. SMN1 produces normal SMN protein, but SMN2 has a mutation that leads to exon 7 deletion, resulting in a short protein that is non-functional. Exon 7 is found in approximately 10–15% of SMN2 mRNA transcripts and allows functional SMN protein to be produced. Spinal muscular atrophy (SMA) patients have a non-functional SMN1 gene and rely on the SMN2 gene to produce the SMN protein, which is a survival requirement. The absence of this protein characterizes SMA, which leads to nerve cells (motor neurons) being selectively degenerated (Malviya et al., 2024). The number of copies of the SMN2 gene correlates with the severity of spinal muscular atrophy (SMA), which was proven by the genotype/phenotype relationship (Blasco-Pérez et al., 2022). However, other phenotypic modifiers can also be discerned, as seen in cases where individuals with two SMN2 copies still had fewer symptoms because of some mutations that increased the synthesis of SMN protein (Costa-Roger et al.,

2021). This heterogeneity proves that SMN2 copy number is not sufficient to diagnose the severity of SMA, so it is important in genetic counselling (Dangouloff et al., 2021). Genetic testing with clinical validation could eventually integrate SMA screening of the newborns, provided the existence of effective therapeutics and an in-depth understanding of genetic modifiers.

4.5.5.6 Clinical Management

4.5.5.6.1 Pulmonary

Respiratory failure is the leading cause of death in infants with type 1 and type 2 Spinal Muscular Atrophy (SMA). It is necessary to consult with a pulmonary specialist who specializes in neuromuscular disorders in children immediately upon the diagnosis (Nishio et al., 2023). The use of a noninvasive ventilatory mode of support is a key factor in increasing survival and life quality in type 1 SMA (Paul et al., 2021). Bi-level positive airway pressure treatment is tolerable and favourable to respiratory functions (Panagiotou et al., 2022). Patients who have damaged respiratory muscles tend to develop aspiration, blockage of the air with mucus, and frequent pulmonary infections. To alleviate the risks, it is recommended that nightly oximetry monitoring be taken in instances of acute illnesses, airway clearance procedures, and pharmaceutical therapies during infections to reduce the associated risks (Nishio et al., 2023). The aim of the pulmonary intervention in type 1 neonates is to sustain a better quality of life and not to extend survival. Non-invasive ventilation has the potential to support or improve chest wall compliance and pulmonary development (Edel et al., 2021). Choices made with regard to any additional processes like tracheostomy and extended ventilatory support are family decisions that require multidisciplinary discussions, including palliative care statements (Panagiotou et al., 2022). The respiratory care of children with type 2 or non-ambulant spinal muscular atrophy has similarities with the respiratory care of newborns with type 1, but with less severe outcomes. Regular measurements of cough effectiveness and respiratory muscle functioning should be undertaken, and in children above the age of five, long-term monitoring of forced vital capacity as well as nocturnal hypoventilation through noninvasive ventilation is recommended (Borrelli et al., 2023).

4.5.5.6.2 Gastrointestinal and Nutritional

Patients with Spinal Muscular Atrophy (SMA) can experience gastrointestinal problems, and these problems could be related to immobility, dietary deficiencies, or underlying gastrointestinal motility issues (Dobkin, 2021). In newborns with type 1 spinal muscular atrophy, long feeding times can represent progressive weakness leading to failure to thrive or aspiration. Gastrointestinal dysfunction cases include bulbar dysfunction and, hence, difficulty in feeding and swallowing, gastrointestinal reflux, sluggish stomach emptying, and constipation (Li et al., 2022). These challenges are not common in ambulatory patients. Dysphagia and aspiration management interventions involve food consistency change to semi-solid and thick liquids, whereas early surgical procedures, e.g. gastrostomy and laparoscopic Nissen fundoplication, are recommended to SMA infants to maintain adequate nutrition and minimize the risk of aspiration-related infections (Corsello et al., 2021). Children with type 2 spinal muscular atrophy are at risk of malnutrition, and it is important to monitor this condition to prevent the degradation of muscle mass. Height and weight clinical measures of patients with SMA can be inaccurate measures of their nutritional status due to reduced lean body mass; non-ambulatory patients are particularly vulnerable to obesity despite low levels of energy intake (Ferrantini et al., 2022). Regular checks conducted by the dietitians are important to ensure there is proper growth and nutrition. In addition, vitamin D and calcium are vital and should be taken in sufficient amounts to counteract the decrease in bone mineral density with age (Liu et al., 2024).

4.5.5.6.3 Orthopaedic and Musculoskeletal Complications

The loss of musculoskeletal functions due to impairment and limited mobility can make it a critical issue, and it is necessary to detect and intervene early to maintain the functionality and quality of life. Stretching and bracing should be regular in non-ambulatory patients with spinal muscular atrophy (SMA) in order to prevent contractures (Oude Lansink et al., 2025). The use of a wheelchair can be introduced as early as 18–24 months, with standing frames being used in the case of patients who can partially bear weight. Physical therapy increases stamina and safety by engaging in swimming and adaptive sports (Vitale et al., 2022). Neuromuscular fatigue may slow down the performance, and the situation has to be moderated by the context to improve accessibility and autonomy. Scoliosis is a frequent instance in non-ambulatory SMA patients and may lead to respiratory complications if otherwise unaddressed; spinal fusion and bracing are considered standard therapies, but their effectiveness is debatable (Brusa et al., 2024; Kolb & Kissel, 2015).

4.5.6 Traumatic Spinal Cord Injury

Traumatic spinal cord injury is a major cause of disability that affects about 54 per million annually, of which over 294,000 have been reported in the United States, and 17,000 additional cases occur each year (National Spinal Cord Injury Statistical Center, 2019). Motor vehicle accidents and falls disproportionately injure young men and non-Hispanic Blacks. The median age of the affected individuals has grown from 29 years in the 1970s to 43 years since 2015 (Chen et al., 2016). This is expected to rise, especially among the older generation. The most common areas affected by spinal injuries are the cervical (50%), the thoracic (35%), and the lumbar (11%) areas. Amidst recent discoveries, patients are faced with reduced life expectancies and high rates of fatalities. The cost of care per year is between $30,770 and $62,563, and lifetime costs amount to between 1.1 million and 4.6 million dollars among young patients (White & Black, 2017). It is important to understand the basic pathophysiology of these injuries to improve patient outcomes. Traumatic spinal cord injury (SCI) normally occurs due to direct injuries to the spine that lead to fracture or dislocation of the vertebrae, exposing the ligaments and interspinal discs. This uproar brings about mechanical pressures resulting in primary injuries, including, but not limited to, contusions, chronic compression of the spinal cord, laceration, and transection, the latter being harder to manage (Wang et al., 2021). Traumatic dislocations or burst fractures cause the most common form of first injury, chronic spinal cord compression. Transient compression could sometimes happen as a result of injuries caused by hyperextension or distraction that causes a separation between two vertebrae. Penetrating injuries are characterized by lacerations and transections with a large dislocation or jagged bone fragments (Ahuja et al., 2017). The primary injury significantly disrupts the spinal cord pathways and blood vessels, leading to an immediate degradation of the blood-spinal cord barrier, vasomotor dysfunction, and increased neuroinflammation. They trigger a cascade of biochemical events that worsen the situation, leading to hypoxia and cell death. Dying neurons release free radicals and disrupt the uptake of glutamate, which increases oxidative damage (Zhang et al., 2021). This step progresses to augmented cell loss of neurons and axonal death due to ongoing neuroinflammatory actions. Accordingly, effective treatment can only be achieved by the early detection of traumatic spinal cord injury, with the primary goal of treatment to prevent or reduce additional damage.

4.5.6.1 Critical Care Management

Severe spinal cord injuries, resulting from acute trauma, can lead to significant respiratory and cardiovascular dysfunction; therefore, urgent medical care is necessary to prevent neurological damage and reduce morbidity. The following are various management strategies.

4.5.6.1.1 Respiratory Management

Respiratory failure may occur due to injuries impacting the brain, cervical spinal cord, or thorax, leading to airway obstruction, aspiration, and impaired diaphragmatic contraction (innervated by C3 to C5) or thoracic accessory muscle function (innervated by T1 to T11). Severe injuries, particularly those with elevated AIS grades and destruction above C3, lead to sudden ventilatory failure. Individuals with partial or mid-cervical injuries may sporadically depend on accessory muscles for respiration; nonetheless, they are susceptible to respiratory failure and necessitate vigilant observation (Hassid et al., 2008). Respiratory difficulties can arise promptly following spinal cord injury, exacerbated by atelectasis, aspiration pneumonia, pulmonary oedema, and mucus accumulation resulting from diminished cough reflexes (Xiang et al., 2021). Respiratory failures may be classified as hypoxemic, hypercapnic, or mixed. In clinical environments, noninvasive ventilation has a restricted function in initial care, rendering prompt orotracheal intubation and suitable laryngoscopy crucial when mechanical ventilation is required. The location of the injury, the extent of neurological damage, and respiratory difficulties influence ventilation in patients with severe spinal cord injuries (Raab et al., 2021). Cervical injuries often require prolonged respiratory support; however, some patients can be successfully weaned, while tracheostomies may be necessary for those with low AIS motor scores, ongoing respiratory issues, and difficulties in managing secretions (Long et al., 2022). Daily ICU therapies, such as respiratory recruitment techniques and chest physiotherapy, improve airway compliance. In instances of chronic respiratory failure, early tracheostomy may reduce the incidence of ventilator-associated pneumonia and decrease ICU length of stay; however, this recommendation is not strongly supported by evidence, and the ideal timing remains unclear (Balas et al., 2023). Preliminary diaphragmatic pacing demonstrates potential in aiding individuals with severe cervical injuries to decrease ventilation length and hospital expenses; nonetheless, additional extensive research is required to provide definitive guidelines (Onders et al., 2022).

4.5.6.1.2 Cardiovascular Management

During the initial phase following traumatic spinal cord injuries, neurogenic shock and hypovolemia may lead to cardiovascular failure, resulting in inadequate spinal cord perfusion and ischemia, which can cause secondary neurological compromise. Early treatment plays an important role in preventing neurological damage and decreasing morbidity. The improvement of mean arterial pressure (MAP) is necessary to maximize the neurological outcome of patients with severe spinal cord injuries, particularly in the first week after the injury (Wecht, 2022). It is recommended that a mean arterial pressure (MAP) be maintained at a range of 85 mm Hg to 90 mm Hg (Kwon et al., 2024). There is evidence that a mean arterial pressure (MAP) above 85 mm Hg during the first 72 hours is correlated with better recovery, but the advantage is less after the first 4 days (Lucci et al., 2021). These MAP goals might require volume resuscitation and careful use of vasopressors, with such factors as bradycardia caused by neurogenic shock taken into account (Inoue et al., 2014). Vasopressors that are likely to be used include dopamine, norepinephrine, and epinephrine, but phenylephrine is not recommended in patients with injuries above T6 because of the possibility of reflex bradycardia. Lower thoracic injuries, which are related to vasodilation, can be treated with phenylephrine.

4.5.6.1.3 Blunt Cerebrovascular Injury Management

The management of blunt cerebrovascular injury is based on criteria such as the cause of injury, its location, severity, the burden of stroke, and the local competence. Neurologists will play a critical role in diagnosing and treating different disorders. Currently, there are no randomized trials that directly compare the antiplatelet and anticoagulant treatments (Das et al., 2022). When starting to use antithrombotic medication, the possibility of the occurrence of haemorrhage should be counterbalanced by the probability of stroke caused by the development of thrombus. As a rule, low-grade injuries (Biffl grades I and II) can be treated with the help of antiplatelet or anticoagulant therapy, and vascular imaging is recommended in case of the appearance of new neurological impairments. The duration of the therapeutic process is usually 3–6 months, followed by outpatient imaging to assess the healing process. The treatment of higher-grade BIFFL injuries (teens and greater), which might need endovascular or surgical procedures, is less effective in the medical intervention.

4.5.6.1.4 Venous Thromboembolism Prevention

Patients who had a severe spinal cord injury are at high risk of developing venous thromboembolism (VTE) during the first 8 weeks after their injuries, which is due to impaired vasomotor tone and limited movement (Godat et al., 2015). The existing recommendations recommend the use of low-molecular-weight heparin within 72 hours after the injury, provided that there is no bleeding; however, not much evidence exists regarding the most beneficial prophylactic method (Daly et al., 2025). The subcutaneous heparin in low doses is ineffective by itself, but it has the potential to provide the advantages of combining with pneumatic compression or electrical stimulation (Chen & Wang, 2013). The recommended duration of thromboprophylaxis is at least 8 weeks because the risks of venous thromboembolism are higher during this period. IUC filters have shown no reduction in the mortality rates; therefore, the use of the filters is limited to specific situations where anticoagulation is contraindicated or not effective (Dhall et al., 2013).

4.5.6.2 Early Recovery and Rehabilitation

The traumatic spinal cord injury significantly reduces the level of independence and physical functioning, which leads to systemic complications such as contractures, rigidity, autonomic dysreflexia, and nutritional deficits, among others. Neurologists play a crucial role in the treatment of such injuries and the complications that they constitute, including autonomic dysfunction, dysphagia, and mental problems. The first recovery begins in critical care that is tailored to the needs of each patient, with regard to the specifics of the injury and the related disorders. The team of therapists, dietitians, and psychologists is also very vital in enhancing the recovery and addressing issues. The major problems during subacute and chronic traumatic spinal cord injury include joint contracture and rigidity. Proper position of the joint is essential to protect the tissue of the joints and maintain the tone of muscles (Hicks et al., 2003). Occupational or physical therapists enhance the range of motion and muscle strength that are built with early physical therapy intervention, suggested in order to preserve muscle retention and counteract deconditioning. The therapy should be done daily for at least 20 minutes. Electrical stimulation can be used to help patients with fatigue in the course of treatment to increase muscle strength and activity (Gill et al., 2018). Moreover, computerized wheelchairs, the neuroprosthetics that are controlled by the brain, enhance functional rehabilitation. More randomized clinical trials are needed to determine the effectiveness of these new techniques on the outcome of recovery.

Autonomic dysreflexia is a complication that occurs during the recovery period of spinal and neurogenic shock, which appears 4–5 days after the injury and may recur repeatedly during the recovery process. It is a disorder caused by unopposed sympathetic stimulation due to stimuli less intense than those that lead to spinal cord injury, the most common precipitants of which are bladder distension and faecal impaction. It is also predominant in a patient with a complete spinal cord injury or a T6 level and above lesion (Eldahan & Rabchevsky, 2018). The symptoms include tachycardia, severe headaches, and alternating hypertension, but the patients may also exhibit flushing and secretions above the lesion and signs of vasoconstriction below the lesion. Treatment underlines the removal of triggering variables and prescribes short-acting direct vasodilators in cases of pharmacotherapy, but it should be done with care to avoid hypotension (Izzy, 2024).

4.5.7 Myasthenia Gravis

Myasthenia gravis is a non-malignant autoimmune disorder characterized by the weakening and exhaustion of the voluntary muscles because of autoantibodies that attack the nicotinic acetylcholine receptor of the neuromuscular junction. It is bimodally distributed, has peaks at the third and sixth decades, and is likely to be underdiagnosed in older groups. Myasthenia gravis is a life-threatening autoimmune disease characterized by the worsening and degeneration of voluntary muscles due to the effects of autoantibodies that attack nicotinic acetylcholine receptors (AChR) at the neuromuscular junction (Remijn-Nelissen et al., 2022). In 1672, Thomas Willis first observed signs of muscle weakness, followed by multiple reports documenting comparable occurrences (Keritam et al., 2024). Initial research suggested that toxic substances were responsible for the damage to lower motor neurons. The identification of acetylcholine as a neurotransmitter resulted in substantial progress in comprehending the pathophysiology and management of the illness (Plomp et al., 2025). The condition is believed to have an autoimmune origin, as evidenced by its correlation with many autoimmune disorders and thymic irregularities (Iyer et al., 2021). Treatment modalities have progressed from neostigmine to corticosteroids, immunosuppressants, and thymectomy, after the identification of AChR antibodies in numerous patients and further antibodies directed against MuSK in a particular fraction (Knezevic et al., 2023). Muscular weakness and fatigability indicate myasthenia gravis, caused by an autoimmune defect on acetylcholine receptors at neuromuscular junctions. Autoantibodies diminish AChR availability via multiple ways. The architecture of the nicotinic acetylcholine receptor is well comprehended, and experimental models have improved understanding of the condition. Substantial progress in diagnosis and treatment has converted myasthenia gravis from a potentially lethal condition to one in which the majority of people may have meaningful and productive lives, due to contemporary immunotherapy. Significant gaps in comprehension remain about the elements that trigger and maintain the autoimmune response in myasthenia gravis, with substantial research being focused on defining these mechanisms.

4.5.7.1 Classifications of Myasthenia Gravis

Myasthenia gravis (MG) may be classified by the age of onset, the presence of anti-AchR antibodies, severity, and aetiology. It is divided into transient neonatal and adult autoimmune. Transient neonatal myasthenia gravis is an illness in infants in which the maternal anti-acetylcholine receptor antibodies cross the placental barrier, and the symptoms of hypotonia and respiratory distress can be detected within a few hours after birth. They tend to relieve the symptoms over 1–3 weeks, which may need supportive treatment or pyridostigmine. Myasthenia gravis may be classified into seropositive and seronegative depending on the presence of antibodies. The most common type is seropositive myasthenia gravis, which affects approximately 85% (in generalized myasthenia gravis) and 50–60% (in ocular myasthenia gravis) of the individuals with anti-AchR antibodies. On the other hand, seronegative (10% to 20%) is expressed by 10–20% of people, or there is no trace of anti-AchR antibodies (Stahl et al., 2025). Some people may have MuSK antibodies, suggesting an immune response that could explain regional muscular weakness and resistance to standard treatment (Dos Santos et al., 2025). Additionally, seronegative patients may display additional humoral variables, including IgG and non-IgG antibodies, that impede AChR activation. The severity of adult myasthenia gravis (MG) is assessed by Osserman's classification, which includes four categories: ocular MG, mild to moderate generalized MG, severe generalized MG, and myasthenic crisis with respiratory failure. Ocular weakness, moderate weakness with or without ocular involvement, and severe weakness with or without established criteria, such as the need for feeding tubes or intubation for respiratory assistance, are the new study categories that have been standardized by a recent classification (Meisel et al., 2024). Myasthenia gravis is characterized by abnormalities at the neuromuscular junction, particularly a reduced quantity of acetylcholine receptors (AChRs), which results in a diminished postsynaptic membrane, a contraction of synaptic folds due to the deterioration of terminal expansions, and an enlargement of synaptic clefts caused by the shortening of junctional folds. These alterations result from an immunological assault on the postsynaptic membrane, in contrast to the presynaptic irregularities observed in Lambert-Eaton syndrome. The reduced safety factor and normal synaptic degeneration cause a progressive attenuation of the amplitude of the end-plate potential (EPP) to a final result of myasthenic weakness and fatigue during exercise.

4.5.7.2 Clinical Features of Myasthenia Gravis

The symptom of muscle fatigue, which is relieved by rest, is generally demonstrated by patients after a long period of exercise. Symptoms differ, and they tend to increase in the evening time and could be aggravated by physical activity, high temperatures, illness, mental stress, certain medications, surgeries, menstruation, and pregnancy. The regularly affected muscles include the levator palpebrae superioris, extraocular, proximal limb, facial expression, and neck extensor muscles, with external ocular muscles implicated initially in some 50% of patients and 90 per cent overall. Ptosis can often be unilateral and heterogeneous (Kasl et al., 2024). Eyelid twitching response is referred to as Cogan lid twitch, and it is an indication of myasthenia gravis. This effect occurs when the patient looks down at a certain period of 10–20 seconds, then returns to the original position, and the eyelid of the eye is raised and then followed by the drooping or twitching. The response shows the rapid recovery and vulnerability to exhaustion associated with the disease. This symptom is not exclusive to myasthenia gravis; it can also be seen in brainstem or eye issues. Ptosis improvement can be observed in sleep or after the rubbing of the eyelid with ice (Shuey, 2022).

Weakness of the orbicularis oculi can result in myasthenia gravis, resulting in difficulties in closing the eyes and uneven and periodically changing ocular palsies, which can mimic the various types of ophthalmoplegia. Saccades are typically hypometric, with a normal velocity at the onset but decreasing as time goes on, contributing to goal undershooting (Azadi et al., 2024). Learners are generally unaffected; however, their facial expressions may seem depressed. This condition can occur because individuals might clench their jaws due to mouth weakness, which can lead to a snarl when they attempt to smile. The symptoms go further to include a nasal voice quality, nasal regurgitation due to palatal weakness, and dysphonia due to laryngeal weakness. Dysphagia is often caused by muscular fatigue at the time of mastication and deglutition (Van Der Heul et al., 2022). Isolated eye weakness (ocular myasthenia) may occur in about 10% of individuals, while the remaining 90% will experience involvement of the facial and limb muscles (generalized myasthenia). The myasthenia gravis progression usually moves onwards to bulbar, truncal, and limb muscles after other muscle groups, such as ocular and facial. Disruption of the intercostal muscles and diaphragm can result in dyspnoea, perhaps presenting as orthopnoea, which improves when sitting up, and the presence of diaphragmatic symmetry, which is an indication of neuromuscular breathlessness (Shuey, 2022). The sudden appearance of dyspnoea requires the attentive monitoring of forced vital capacity. Major conditions might require intubation and ventilators. Exertional fatigue is observed in the patients, which is explained by the involvement of limb muscles, but deep tendon reflexes are normal or increased, and the patient does not have sensory symptoms. The fluctuation of weakness and spontaneous remissions, particularly in the initial phases, hinders objective evaluation; still, total remissions are rare (Radha & Lopus, 2021; Thanvi & Lo, 2004).

4.5.8 Amyotrophic Lateral Sclerosis

Amyotrophic lateral sclerosis (ALS) is a rare and lethal neurodegenerative disorder of the central nervous system, frequently difficult to diagnose in the early stages due to the focus on more prevalent disorders in the differential diagnosis. Amyotrophic lateral sclerosis (ALS) is defined by the degeneration of upper and lower motor neurons across several regions, resulting in the gradual atrophy of voluntary muscles involved in limb movement, swallowing, speech, and respiration (Goutman et al., 2022a). Sphincter and extraocular muscles are often preserved, however autonomic dysfunction, characterized by urine urgency, is observed. Clinical weakness typically disseminates contralaterally and throughout neighbouring anatomical segments, often initiating with a localized onset in a particular region (Walhout et al., 2018). The principal kinds of ALS include bulbar and spinal onset, in addition to less common varieties such as flail arm and leg syndrome, and primary lateral sclerosis. Phenotypes depend on age, sex, and genetics, which proves their variability across the world (Chiò et al., 2020). The age of onset is generally older, and the bulbar onset is more common in German ALS patients compared to their Chinese counterparts. A better consensus of phenotyping is instrumental in understanding the way demographic factors affect ALS symptoms (Rosenbohm et al., 2018). The previous understanding of Amyotrophic Lateral Sclerosis (ALS) has gone a step further to derive cognitive and behavioural changes, which are observed in 35 to 50% of patients (Pender et al., 2020). One notable symptom is motor dysfunction, but individuals may also be impaired in terms of language and executive functions, such as impaired working memory and inhibition; long-term memory and spatial functions are generally unaffected. The changes in behaviour might include apathy, impatience, and lack of attention to personal cleanliness, as well as changes in eating habits, and about 15% of them will meet the standards of frontotemporal dementia (FTD). Additionally, patients with ALS could suffer sorrow, anxiety, and sleep disorders, as well as pseudobulbar affect, which also leads to emotional instability (Ringholz et al., 2005).

The changes in cognition and behaviour of ALS suggest that it is a complicated neurodegenerative disease linked to FTD with TDP-43 proteinopathy exhibited in nearly 97,000 cases of ALS and 50,000 cases of FTD. The presence of deficits in executive function, language, and fluency demonstrates flawless specificity of TDP-43 pathology in the relevant areas of the brain. Patient factors, including C9orf72 status and onset of bulbar symptoms, are essential predictors of cognitive impairment, which implies that early identification of these features can be used to predict the occurrence of problems that develop (Iazzolino et al., 2021). Expedited disease progression and lowered survival rates are linked to cognitive deterioration, observed in 30% of ALS patients in 6 months (Bersano et al., 2020). According to MRI studies, the disruption of the motor and non-motor networks is significant, and the structural connectivity impairment is related to the motor impairments, whereas the functional connectivity issue is connected to cognitive and behavioural changes (Basaia et al., 2020). This understanding highlights the importance of treating the cognitive impairment and associated neuropsychological conditions, including depression and sleep disorders. In cases where cognitive symptoms occur, healthcare providers need to engage patients and families in end-of-life preference conversations, which will necessitate active participation (Nicholson et al., 2018). It is important to identify factors that predispose and promote the progression of ALS to treat the patients. Genetics, especially mutations such as C9orf72 expansions, are elements of high risk that influence the onset and the survival rates (Westeneng et al., 2018). Certain genetic deviations only affect risk exposure but not the course of the disease. The exposome, which includes all environmental exposures a patient has experienced throughout their lifetime, is associated with the development of ALS and may also cause its progression. The pathophysiology of ALS forms the basis of developing new therapeutic and preventative interventions and genetic therapies for carriers of asymptomatic mutations (Goutman et al., 2019).

4.5.8.1 Molecular Pathomechanisms in ALS

The mutations of more than 40 known ALS genes lead to toxic effects and protein aggregation, which are the pathogenic processes in ALS. Pathophysiology processes can be categorized as disturbed RNA metabolism, altered proteostasis/autophagy, cytoskeletal/trafficking, and mitochondrial dysfunction. The interference of RNA metabolism occurs through the critical genes such as C9orf72, TARDBP, and FUS, which lead to the accumulation of TDP-43 and FUS proteins, therefore blocking transcription and RNA processing. By blocking the elimination of damaged proteins by promoting proteostasis and autophagy, TDP-43 alters proteostasis and autophagy, and mutations in TUBA4A and PFN1 cause cytoskeletal defects blocking axonal transport. Oxidative stress is further worsened by mitochondrial dysfunction, particularly caused by SOD1 mutations, and therefore, its role in the aetiology of ALS is significant.

Despite significant advances, the pathophysiology of amyotrophic lateral sclerosis (ALS) is poorly understood on a complex molecular basis. The key processes include the PrP-type spreading of the TDP-43 and SOD1 aggregates, which could support the spread of ALS disease (Pokrishevsky et al., 2016). Such genes as TARDBP and FUS are associated with impaired DNA repair systems in ALS; namely, the reduction of nuclear TDP-43 causes the formation of double-stranded DNA breaks, which affects genomic stability (Freibaum et al., 2015). Nucleocytoplasmic transport is disrupted by TDP-43 aggregates, mutant FUS, and C9orf72 repeat expansions (Chou et al., 2018). Furthermore, neurotoxic dipeptide repeat proteins of C9orf72 expansion transcripts might cause abnormalities of heterochromatin formation and facilitate TDP-43 aggregation (Cook et al., 2020). The role of central and peripheral inflammatory pathways on genetic changes and progression of Amyotrophic Lateral Sclerosis (ALS), which has a sporadic variant, is substantive (McCauley et al., 2020). In ALS, the number of immune cells, the state of their activation, and the production of cytokines change (Murdock et al., 2017). The immune system follows a complex mechanism initially offering resistance, but eventually resulting in a harmful cytotoxic response. Furthermore, hypermetabolic nature is a common symptom of ALS, with genetic changes and the general progression of the illness. Metabolomic studies can reveal molecular changes that can lead to the development of illnesses (Goutman et al., 2022b; Feldman et al., 2022). As part of developing new therapeutic approaches and prevention interventions for ALS, it is necessary to understand the mechanisms of the connection between ALS genes, inflammation, and hypermetabolism.

4.6 Conclusion

Overall, muscle and motor system disabilities are a major cause of functional impairment, which impairs mobility, coordination, strength, and the capacity to perform daily activities without help. Muscular dystrophies, myopathy, peripheral neuropathy, post-polio syndrome, spinal muscular atrophy, traumatic spinal cord injury, myasthenia gravis, and amyotrophic lateral sclerosis are some examples of causes and clinical manifestations of motor impairments. The disorders may be caused by genetic factors, neuromuscular degeneration, nerve damage, or traumatic injuries, often leading to progressive physical decline and the need for long-term care. Early evaluation, correct diagnosis, and prompt rehabilitation interventions are crucial in preserving motor functions, minimizing complications, and enhancing overall quality of life. Multidisciplinary care, physiotherapy, and assistive devices are important in the promotion of functional independence and involvement in daily activities. On top of physical issues, people with muscle and motor system disabilities often encounter social and environmental obstacles, such as the inaccessibility of the physical environment, reliance on caregivers, exclusion from physical and recreational activities, and restricted access to education and employment. Thus, medical management, rehabilitation, and social support are to be used to treat people with muscle and motor system disabilities holistically to ensure their dignity, independence, and social inclusion.

References

Abdo, W. F., Van De Warrenburg, B. P., Burn, D. J., Quinn, N. P., & Bloem, B. R. (2010, January). The clinical approach to movement disorders. *Nature Reviews Neurology, 6*(1), 29–37.

Ahuja, C. S., Nori, S., Tetreault, L., Wilson, J., Kwon, B., Harrop, J., Choi, D., & Fehlings, M. G. (2017, March 1). Traumatic spinal cord injury—Repair and regeneration. *Neurosurgery, 80*(3S), S9–S22.

Aljarallah, S., Alkhawajah, N., Aldosari, O., Alhuqbani, M., Alqifari, F., Alkhuwaitir, B., Aldawood, A., Alshenawy, O., & BaHammam, A. S. (2023, June 19). Restless leg syndrome in multiple sclerosis: A case–control study. *Frontiers in Neurology, 14*, 1194212.

Annoussamy, M., Seferian, A. M., Daron, A., Péréon, Y., Cances, C., Vuillerot, C., De Waele, L., Laugel, V., Schara, U., Gidaro, T., & Lilien, C. (2021, February). Natural history of type 2 and 3 spinal muscular atrophy: 2-year NatHis-SMA study. *Annals of Clinical and Translational Neurology, 8*(2), 359–373.

Azadi, R., Holcombe, A. O., & Edelman, J. A. (2024, January 2). Hypometria of saccadic eye movements to targets in rapid circular motion. *Journal of Vision, 24*(1), 2.

Balas, M., Jaja, B. N., Harrington, E. M., Jack, A. S., Hofereiter, J., Malhotra, A. K., Jaffe, R. H., He, Y., Byrne, J. P., Wilson, J. R., & Witiw, C. D. (2023, December 1). Earlier tracheostomy reduces complications in complete cervical spinal cord injury in real-world practice: Analysis of a multicenter cohort of 2001 patients. *Neurosurgery, 93*(6), 1305–1312.

Basaia, S., Agosta, F., Cividini, C., Trojsi, F., Riva, N., Spinelli, E. G., Moglia, C., Femiano, C., Castelnovo, V., Canu, E., & Falzone, Y. (2020, November 3). Structural and functional brain connectome in motor neuron diseases: A multicenter MRI study. *Neurology, 95*(18), e2552–e2564.

Battini, R., Chieffo, D., Bulgheroni, S., Piccini, G., Pecini, C., Lucibello, S., Lenzi, S., Moriconi, F., Pane, M., Astrea, G., & Baranello, G. (2018, February 1). Cognitive profile in Duchenne muscular dystrophy boys without intellectual disability: The role of executive functions. *Neuromuscular Disorders, 28*(2), 122–128.

Benatar, M., Wuu, J., McHutchison, C., Postuma, R. B., Boeve, B. F., Petersen, R., Ross, C. A., Rosen, H., Arias, J. J., Fradette, S., & McDermott, M. P. (2022, January 1). Preventing amyotrophic lateral sclerosis: Insights from pre-symptomatic neurodegenerative diseases. *Brain, 145*(1), 27–44.

Benson, M. D., Waddington-Cruz, M., Berk, J. L., Polydefkis, M., Dyck, P. J., Wang, A. K., Planté-Bordeneuve, V., Barroso, F. A., Merlini, G., Obici, L., & Scheinberg, M. (2018, July 5). Inotersen treatment for patients with hereditary transthyretin amyloidosis. *New England Journal of Medicine, 379*(1), 22–31.

Berger, A. R., Schaumburg, H. H., Schroeder, C., Apfel, S., & Reynolds, H. (1992, July). Dose response, coasting, and differential fibre vulnerability in human toxic neuropathy: A prospective study of pyridoxine neurotoxicity. *Neurology, 42*(7), 1367.

Bersano, E., Sarnelli, M. F., Solara, V., Iazzolino, B., Peotta, L., De Marchi, F., Facchin, A., Moglia, C., Canosa, A., Calvo, A., & Chiò, A. (2020, July 2). Decline of cognitive and behavioral functions in amyotrophic lateral sclerosis: A longitudinal study. *Amyotrophic Lateral Sclerosis and Frontotemporal Degeneration, 21*(5–6), 373–379.

Bertrand, A. T., Chikhaoui, K., Yaou, R. B., & Bonne, G. (2011, December 1). Clinical and genetic heterogeneity in laminopathies. *Biochemical Society Transactions, 39*(6), 1687–1692.

Blasco-Pérez, L., Costa-Roger, M., Leno-Colorado, J., Bernal, S., Alias, L., Codina-Solà, M., Martínez-Cruz, D., Castiglioni, C., Bertini, E., Travaglini, L., & Millán, J. M. (2022, July 27). Deep molecular characterization of milder spinal muscular atrophy patients carrying the c.859G>C variant in SMN2. *International Journal of Molecular Sciences, 23*(15), 8289.

Bloem, B. R., Okun, M. S., & Klein, C. (2021, June 12). Parkinson's disease. *The Lancet, 397*(10291), 2284–2303.

Bögershausen, N., Shahrzad, N., Chong, J. X., von Kleist-Retzow, J. C., Stanga, D., Li, Y., Bernier, F. P., Loucks, C. M., Wirth, R., Puffenberger, E. G., & Hegele, R. A. (2013, July 11). Recessive TRAPPC11 mutations cause a disease spectrum of limb girdle muscular dystrophy and myopathy with movement disorder and intellectual disability. *The American Journal of Human Genetics, 93*(1), 181–190.

Bolduc, V., Marlow, G., Boycott, K. M., Saleki, K., Inoue, H., Kroon, J., Itakura, M., Robitaille, Y., Parent, L., Baas, F., & Mizuta, K. (2010, February 12). Recessive mutations in the putative calcium-activated chloride channel Anoctamin 5 cause proximal LGMD2L and distal MMD3 muscular dystrophies. *The American Journal of Human Genetics, 86*(2), 213–221.

Bolduc, V., Foley, A. R., Solomon-Degefa, H., Sarathy, A., Donkervoort, S., Hu, Y., Chen, G. S., Sizov, K., Nalls, M., Zhou, H., & Aguti, S. (2019, March 21). A recurrent COL6A1 pseudoexon insertion causes muscular dystrophy and is effectively targeted by splice-correction therapies. *JCI Insight, 4*(6), e124403.

Bonne, G., Mercuri, E., Muchir, A., Urtizberea, A., Becane, H. M., Recan, D., Merlini, L., Wehnert, M., Boor, R., Reuner, U., & Vorgerd, M. (2000, August). Clinical and molecular genetic spectrum of autosomal dominant Emery-Dreifuss muscular dystrophy due to mutations of the lamin A/C gene. *Annals of Neurology: Official Journal of the American Neurological Association and the Child Neurology Society, 48*(2), 170–180.

Borrelli, M., Terrone, G., Evangelisti, R., Fedele, F., Corcione, A., & Santamaria, F. (2023, March 1). Respiratory phenotypes of neuromuscular diseases: A challenging issue for pediatricians. *Pediatrics & Neonatology, 64*(2), 109–118.

Brusa, C., Baranello, G., Ridout, D., de Graaf, J., Manzur, A. Y., Munot, P., Sarkozy, A., Main, M., Milev, E., Iodice, M., & Ramsey, D. (2024, November). Secondary outcomes of scoliosis surgery in disease-modifying treatment-naïve patients with spinal muscular atrophy type 2 and nonambulant type 3. *Muscle & Nerve, 70*(5), 1000–1009.

Callaghan, B. C., Little, A. A., Feldman, E. L., & Hughes, R. A. (2012). Enhanced glucose control for preventing and treating diabetic neuropathy. *Cochrane Database of Systematic Reviews, 6.*

Cances, C., Vlodavets, D., Comi, G. P., Masson, R., Mazurkiewicz-Bełdzińska, M., Saito, K., Zanoteli, E., Dodman, A., El-Khairi, M., Gorni, K., & Gravestock, I. (2022, July 29). Natural history of type 1 spinal muscular atrophy: A retrospective, global, multicenter study. *Orphanet Journal of Rare Diseases, 17*(1), 300.

Cardamone, M., Darras, B. T., & Ryan, M. M. (2008, April). Inherited myopathies and muscular dystrophies. *Seminars in Neurology, 28*(02), 250–259.

Cheeran, D., Khan, S., Khera, R., Bhatt, A., Garg, S., Grodin, J. L., Morlend, R., Araj, F. G., Amin, A. A., Thibodeau, J. T., & Das, S. (2017, October 17). Predictors of death in adults with Duchenne muscular dystrophy–associated cardiomyopathy. *Journal of the American Heart Association, 6*(10), e006340.

Chen, H. L., & Wang, X. D. (2013, August). Heparin for venous thromboembolism prophylaxis in patients with acute spinal cord injury: A systematic review and meta-analysis. *Spinal Cord, 51*(8), 596–602.

Chen, Y., He, Y., & DeVivo, M. J. (2016, October 1). Changing demographics and injury profile of new traumatic spinal cord injuries in the United States, 1972–2014. *Archives of Physical Medicine and Rehabilitation, 97*(10), 1610–1619.

Chiò, A., Moglia, C., Canosa, A., Manera, U., D'Ovidio, F., Vasta, R., Grassano, M., Brunetti, M., Barberis, M., Corrado, L., & D'Alfonso, S. (2020, February 25). ALS phenotype is influenced by age, sex, and genetics: A population-based study. *Neurology, 94*(8), e802–e810.

Chou, C. C., Zhang, Y. I., Umoh, M. E., Vaughan, S. W., Lorenzini, I., Liu, F., Sayegh, M., Donlin-Asp, P. G., Chen, Y. H., Duong, D. M., & Seyfried, N. T. (2018, February). TDP-43 pathology disrupts nuclear pore complexes and nucleocytoplasmic transport in ALS/FTD. *Nature Neuroscience, 21*(2), 228–239.

Christidi, F., Karavasilis, E., Argyropoulos, G. D., Velonakis, G., Zouvelou, V., Murad, A., Evdokimidis, I., Rentzos, M., Seimenis, I., & Bede, P. (2022, April 24). Neurometabolic alterations in motor neuron disease: Insights from magnetic resonance spectroscopy. *Journal of Integrative Neuroscience, 21*(3), 87.

Coelho, T., Marques, W., Dasgupta, N. R., Chao, C. C., Parman, Y., França, M. C., Guo, Y. C., Wixner, J., Ro, L. S., Calandra, C. R., & Kowacs, P. A. (2023, October 17). Eplontersen for hereditary transthyretin amyloidosis with polyneuropathy. *Journal of the American Medical Association, 330*(15), 1448–1458.

Connolly, A., Bailey, S., Lamont, R., & Tu, A. (2024, August 17). Factors associated with assistive technology prescription and acceptance in motor neurone disease. *Disability and Rehabilitation: Assistive Technology, 19*(6), 2229–2238.

Conte, A., Defazio, G., Ferrazzano, G., Hallett, M., Macerollo, A., Fabbrini, G., & Berardelli, A. (2013, June 11). Is increased blinking a form of blepharospasm? *Neurology, 80*(24), 2236–2241.

Cook, C. N., Wu, Y., Odeh, H. M., Gendron, T. F., Jansen-West, K., Del Rosso, G., Yue, M., Jiang, P., Gomes, E., Tong, J., & Daughrity, L. M. (2020, September 2). C9orf72 poly (GR) aggregation induces TDP-43 proteinopathy. *Science Translational Medicine, 12*(559), eabb3774.

Corsello, A., Scatigno, L., Pascuzzi, M. C., Calcaterra, V., Dilillo, D., Vizzuso, S., Pelizzo, G., Zoia, E., Mandelli, A., Govoni, A., & Bosetti, A. (2021, July 13). Nutritional, gastrointestinal and endo-metabolic challenges in the management of children with spinal muscular atrophy type 1. *Nutrients, 13*(7), 2400.

Cortez, M., Singleton, J. R., & Smith, A. G. (2014, January 1). Glucose intolerance, metabolic syndrome, and neuropathy. *Handbook of Clinical Neurology, 126*, 109–122.

Costa-Roger, M., Blasco-Pérez, L., Cuscó, I., & Tizzano, E. F. (2021, August 21). The importance of digging into the genetics of SMN genes in the therapeutic scenario of spinal muscular atrophy. *International Journal of Molecular Sciences, 22*(16), 9029.

Cousins, S. (2017, April 15). Accounting for polio survivors in the post-polio world. *The Lancet, 389*(10078), 1503–1504.

Currò, R., Salvalaggio, A., Tozza, S., Gemelli, C., Dominik, N., Galassi Deforie, V., Magrinelli, F., Castellani, F., Vegezzi, E., Businaro, P., & Callegari, I. (2021, May 1). RFC1 expansions are a common cause of idiopathic sensory neuropathy. *Brain, 144*(5), 1542–1550.

Daly, J. E., Smith, R., Li, C., & Sippel, J. L. (2025, July 11). The history of clinical services, medical training, and research for spinal cord injuries and disorders in the United States veterans health administration. *Archives of Rehabilitation Research and Clinical Translation*, 100488.

Dangouloff, T., Vrščaj, E., Servais, L., Osredkar, D., Adoukonou, T., Aryani, O., Barisic, N., Bashiri, F., Bastaki, L., Benitto, A., & Omran, T. B. (2021, June 1). Newborn screening programs for spinal muscular atrophy worldwide: Where we stand and where to go. *Neuromuscular Disorders, 31*(6), 574–582.

Das, A. S., Vicenty-Padilla, J. C., Chua, M. M., Jeelani, Y., Snider, S. B., Regenhardt, R. W., Al-Mufti, F., Du, R., & Izzy, S. (2022, December 1). Cerebrovascular injuries in traumatic brain injury. *Clinical Neurology and Neurosurgery, 223*, 107479.

de Haan, W. (2010, May 1). Lipodystrophy and muscular dystrophy caused by PTRF mutations. *Clinical Genetics, 77*(5), 436.

Demir, E., Sabatelli, P., Allamand, V., Ferreiro, A., Moghadaszadeh, B., Makrelouf, M., Topaloglu, H., Echenne, B., Merlini, L., & Guicheney, P. (2002, June 1). Mutations in COL6A3 cause severe and mild phenotypes of Ullrich congenital muscular dystrophy. *The American Journal of Human Genetics, 70*(6), 1446–1458.

Dhall, S. S., Hadley, M. N., Aarabi, B., Gelb, D. E., Hurlbert, R. J., Rozzelle, C. J., Ryken, T. C., Theodore, N., & Walters, B. C. (2013, March 1). Deep venous thrombosis and thromboembolism in patients with cervical spinal cord injuries. *Neurosurgery, 72*, 244–254.

Ding, Q., Li, X., Wang, M., Wang, J., Sun, T., Sun, Y., Liu, J., Yan, Y., Wu, J., Du, J., & Dong, X. (2025, October 10). Obstructive sleep apnea in community-dwelling polio survivors: A 5-year longitudinal follow-up study. *Frontiers in Neurology, 16*, 1643862.

Dobkin, B. H. (2021, March 23). Paraplegia and spinal cord syndromes. In *Bradley's neurology in clinical practice E-Book* (pp. 356–361). Elsevier.

Dos Santos, R. X., Waelkens, J., Crawford, A. H., Khan, S., Sami, S., Gomes, S. A., Van Ham, A., Van Soens, I., Cornelis, I., Canning, J., & Fenn, J. (2025, May). Case series of canine myasthenia gravis: A classification approach with consideration of seronegative dogs. *Journal of Veterinary Internal Medicine, 39*(3), e70113.

Dunaway Young, S., McGrattan, K., Johnson, E., Van Der Heul, M., Duong, T., Bakke, M., Werlauff, U., Pasternak, A., Cattaneo, C., Hoffman, K., & Fanelli, L. (2023, July 4). Development of an international SMA bulbar assessment for inter-professional administration. *Journal of Neuromuscular Diseases, 10*(4), 639–652.

Dyck, P. J., Oviatt, K. F., & Lambert, E. H. (1981, September). Intensive evaluation of referred unclassified neuropathies yields improved diagnosis. *Annals of Neurology: Official Journal of the American Neurological Association and the Child Neurology Society, 10*(3), 222–226.

Dyck, P. J., Kratz, K. M., Karnes, J. L., Litchy, W. J., Klein, R., Pach, J. M., Wilson, D. M., O'brien, P. C., & Melton, I. L. J. (1993, April). The prevalence by staged severity of various types of diabetic neuropathy, retinopathy, and nephropathy in a population-based cohort: The Rochester Diabetic Neuropathy Study. *Neurology, 43*(4), 817.

Dyck, P. J., Norell, J. E., & Dyck, P. J. (1999, December 1). Microvasculitis and ischemia in diabetic lumbosacral radiculoplexus neuropathy. *Neurology, 53*(9), 2113.

Edel, L., Grime, C., Robinson, V., Manzur, A., Abel, F., Munot, P., Ridout, D., Scoto, M., Muntoni, F., & Chan, E. (2021, April 1). A new respiratory scoring system for evaluation of respiratory outcomes in children with spinal muscular atrophy type1 (SMA1) on SMN enhancing drugs. *Neuromuscular Disorders, 31*(4), 300–309.

Eldahan, K. C., & Rabchevsky, A. G. (2018, January 1). Autonomic dysreflexia after spinal cord injury: Systemic pathophysiology and methods of management. *Autonomic Neuroscience, 209*, 59–70.

Elsayed, N. A., Aleppo, G., Bannuru, R. R., Bruemmer, D., Collins, B. S., Ekhlaspour, L., Gibbons, C. H., Giurini, J. M., Hilliard, M. E., Johnson, E. L., & Khunti, K. (2024, January 2). 12. Retinopathy, neuropathy, and foot care: Standards of care in diabetes—2024. *Diabetes Care, 47*, S231.

Emery, A. E. (2002, February 23). The muscular dystrophies. *The Lancet, 359*(9307), 687–695.

Erokhina, E. K., Melnik, E. A., Lebedeva, D. D., Shamtieva, K. V., Peters, T. V., Pavlikova, E. P., Gepard, V. V., & Vlodavets, D. V. (2024, January). Sleep disorders and fatigue in patients with different forms of myotonic dystrophy type 1. *Neuroscience and Behavioral Physiology, 54*(1), 35–40.

Farbu, E., Gilhus, N. E., Barnes, M. P., Borg, K., De Visser, M., Driessen, A., Howard, R., Nollet, F., Opara, J., & Stalberg, E. (2006, August). EFNS guideline on diagnosis and management of post-polio syndrome. Report of an EFNS task force. *European Journal of Neurology, 13*(8), 795–801.

Feldman, E. L., Russell, J. W., Löscher, W. N., Grisold, W., & Meng, S. (2021, February 25). Motor neuron diseases. In *Atlas of neuromuscular diseases: A practical guideline* (pp. 313–320). Springer.

Feldman, E. L., Goutman, S. A., Petri, S., Mazzini, L., Savelieff, M. G., Shaw, P. J., & Sobue, G. (2022, October 15). Amyotrophic lateral sclerosis. *The Lancet, 400*(10360), 1363–1380.

Ferrantini, G., Coratti, G., Onesimo, R., Lucibello, S., Bompard, S., Turrini, I., Cicala, G., Caprarelli, M., Pera, M. C., Bravetti, C., & Berti, B. (2022, May). Body mass index in type 2 spinal muscular atrophy: A longitudinal study. *European Journal of Pediatrics, 181*(5), 1923–1932.

Finkel, R. S., McDermott, M. P., Kaufmann, P., Darras, B. T., Chung, W. K., Sproule, D. M., Kang, P. B., Foley, A. R., Yang, M. L., Martens, W. B., & Oskoui, M. (2014, August 26). Observational study of spinal muscular atrophy type I and implications for clinical trials. *Neurology, 83*(9), 810–817.

Foley, A. R., Hu, Y., Zou, Y., Yang, M., Medne, L., Leach, M., Conlin, L. K., Spinner, N., Shaikh, T. H., Falk, M., & Neumeyer, A. M. (2011, January). Large genomic deletions: A novel cause of Ullrich congenital muscular dystrophy. *Annals of Neurology, 69*(1), 206–211.

Freibaum, B. D., Lu, Y., Lopez-Gonzalez, R., Kim, N. C., Almeida, S., Lee, K. H., Badders, N., Valentine, M., Miller, B. L., Wong, P. C., & Petrucelli, L. (2015, September 3). GGGGCC repeat expansion in C9orf72 compromises nucleocytoplasmic transport. *Nature, 525*(7567), 129–133.

Gibbons, C. H., & Freeman, R. (2015, January 1). Treatment-induced neuropathy of diabetes: An acute, iatrogenic complication of diabetes. *Brain, 138*(1), 43–52.

Gill, M. L., Grahn, P. J., Calvert, J. S., Linde, M. B., Lavrov, I. A., Strommen, J. A., Beck, L. A., Sayenko, D. G., Van Straaten, M. G., Drubach, D. I., & Veith, D. D. (2018, November).

Neuromodulation of lumbosacral spinal networks enables independent stepping after complete paraplegia. *Nature Medicine, 24*(11), 1677–1682.

Godat, L. N., Kobayashi, L., Chang, D. C., & Coimbra, R. (2015, March 1). Can we ever stop worrying about venous thromboembolism after trauma? *Journal of Trauma and Acute Care Surgery, 78*(3), 475–481.

Godfrey, C., Foley, A. R., Clement, E., & Muntoni, F. (2011, June 1). Dystroglycanopathies: Coming into focus. *Current Opinion in Genetics & Development, 21*(3), 278–285.

Goemans, N., Klingels, K., Van den Hauwe, M., Boons, S., Verstraete, L., Peeters, C., Feys, H., & Buyse, G. (2013, December 31). Six-minute walk test: Reference values and prediction equation in healthy boys aged 5 to12 years. *PLoS One, 8*(12), e84120.

Goselink, R. J., Schreuder, T. H., Mul, K., Voermans, N. C., Pelsma, M., de Groot, I. J., van Alfen, N., Franck, B., Theelen, T., Lemmers, R. J., & Mah, J. K. (2016, August 17). Facioscapulohumeral dystrophy in children: Design of a prospective, observational study on natural history, predictors and clinical impact (iFocus FSHD). *BMC Neurology, 16*(1), 138.

Goutman, S. A., Boss, J., Patterson, A., Mukherjee, B., Batterman, S., & Feldman, E. L. (2019, August 1). High plasma concentrations of organic pollutants negatively impact survival in amyotrophic lateral sclerosis. *Journal of Neurology, Neurosurgery & Psychiatry, 90*(8), 907–912.

Goutman, S. A., Hardiman, O., Al-Chalabi, A., Chió, A., Savelieff, M. G., Kiernan, M. C., & Feldman, E. L. (2022a, May 1). Recent advances in the diagnosis and prognosis of amyotrophic lateral sclerosis. *The Lancet Neurology, 21*(5), 480–493.

Goutman, S. A., Guo, K., Savelieff, M. G., Patterson, A., Sakowski, S. A., Habra, H., Karnovsky, A., Hur, J., & Feldman, E. L. (2022b, December 1). Metabolomics identifies shared lipid pathways in independent amyotrophic lateral sclerosis cohorts. *Brain, 145*(12), 4425–4439.

Hanewinckel, R., van Oijen, M., Ikram, M. A., & van Doorn, P. A. (2016, January). The epidemiology and risk factors of chronic polyneuropathy. *European Journal of Epidemiology, 31*(1), 5–20.

Hassid, V. J., Schinco, M. A., Tepas, J. J., Griffen, M. M., Murphy, T. L., Frykberg, E. R., & Kerwin, A. J. (2008, December 1). Definitive establishment of airway control is critical for optimal outcome in lower cervical spinal cord injury. *Journal of Trauma and Acute Care Surgery, 65*(6), 1328–1332.

Hattori, N., Ichimura, M., Nagamatsu, M., Li, M., Yamamoto, K., Kumazawa, K., Mitsuma, T., & Sobue, G. (1999, March 1). Clinicopathological features of Churg–Strauss syndrome-associated neuropathy. *Brain, 122*(3), 427–439.

Hicks, A. L., Martin, K. A., Ditor, D. S., Latimer, A. E., Craven, C., Bugaresti, J., & McCartney, N. (2003, January). Long-term exercise training in persons with spinal cord injury: Effects on strength, arm ergometry performance and psychological well-being. *Spinal Cord, 41*(1), 34–43.

Hoffman, E. M., Staff, N. P., Robb, J. M., St. Sauver, J. L., Dyck, P. J., & Klein, C. J. (2015, April 21). Impairments and comorbidities of polyneuropathy revealed by population-based analyses. *Neurology, 84*(16), 1644–1651.

Huang, K., Huang, Y., Fang, X., Zhang, Z., Wang, Y., & Peng, D. (2024, December). Elucidation of the gene regulatory network related to spinal muscular atrophy. *Cytology and Genetics, 58*(6), 615–626.

Iazzolino, B., Peotta, L., Zucchetti, J. P., Canosa, A., Manera, U., Vasta, R., Grassano, M., Palumbo, F., Brunetti, M., Barberis, M., & Sbaiz, L. (2021, January 5). Differential neuropsychological profile of patients with amyotrophic lateral sclerosis with and without C9orf72 mutation. *Neurology, 96*(1), e141–e152.

Inoue, T., Manley, G. T., Patel, N., & Whetstone, W. D. (2014, February 1). Medical and surgical management after spinal cord injury: Vasopressor usage, early surgerys, and complications. *Journal of Neurotrauma, 31*(3), 284–291.

Iyer, S. R., Shah, S. B., & Lovering, R. M. (2021, July 28). The neuromuscular junction: Roles in aging and neuromuscular disease. *International Journal of Molecular Sciences, 22*(15), 8058.

Izzy, S. (2024, February 1). Traumatic spinal cord injury. *Continuum: Lifelong Learning in Neurology, 30*(1), 53–72.

Jablonka, S., Hennlein, L., & Sendtner, M. (2022, January 4). Therapy development for spinal muscular atrophy: Perspectives for muscular dystrophies and neurodegenerative disorders. *Neurological Research and Practice, 4*(1), 2.

Jellinger, K. A. (2025, June). Co-occurrence of amyotrophic lateral sclerosis and multiple sclerosis: A rare but interesting association. *Journal of Neural Transmission, 28*, 1–3.

Jimenez-Mallebrera, C., Torelli, S., Feng, L., Kim, J., Godfrey, C., Clement, E., Mein, R., Abbs, S., Brown, S. C., Campbell, K. P., & Kröger, S. (2009, October). A comparative study of α-dystroglycan glycosylation in dystroglycanopathies suggests that the hypoglycosylation of α-dystroglycan does not consistently correlate with clinical severity. *Brain Pathology, 19*(4), 596–611.

Kaplan, J. C., & Hamroun, D. (2015, December 1). The 2016 version of the gene table of monogenic neuromuscular disorders (nuclear genome). *Neuromuscular Disorders, 25*(12), 991–1020.

Kasl, Z., Poczos, P., Herzig, R., Jiraskova, N., Matuska, M., & Cesak, T. (2024, June 5). Neurological disorders. In *Ocular manifestations of systemic diseases* (pp. 375–446). Springer.

Katzberg, H. D., & Kassardjian, C. D. (2016, December). Toxic and endocrine myopathies. *Continuum, 22*(6), 1815–1828.

Kaufmann, P., McDermott, M. P., Darras, B. T., Finkel, R., Kang, P., Oskoui, M., Constantinescu, A., Sproule, D. M., Foley, A. R., Yang, M., & Tawil, R. (2011, June 13). Observational study of spinal muscular atrophy type 2 and 3: Functional outcomes over 1 year. *Archives of Neurology, 68*(6), 779–786.

Keinath, M. C., Prior, D. E., & Prior, T. W. (2021, January 25). Spinal muscular atrophy: Mutations, testing, and clinical relevance. *The Application of Clinical Genetics, 14*, 11–25.

Keritam, O., Vincent, A., Zimprich, F., & Cetin, H. (2024, December 5). A clinical perspective on muscle specific kinase antibody positive myasthenia gravis. *Frontiers in Immunology, 15*, 1502480.

Knezevic, E., Nenic, K., Milanovic, V., & Knezevic, N. N. (2023, November 29). The role of cortisol in chronic stress, neurodegenerative diseases, and psychological disorders. *Cells, 12*(23), 2726.

Koike, H., Ito, S., Morozumi, S., Kawagashira, Y., Iijima, M., Hattori, N., Tanaka, F., & Sobue, G. (2008, July 1). Rapidly developing weakness mimicking Guillain-Barré syndrome in beriberi neuropathy: Two case reports. *Nutrition, 24*(7–8), 776–780.

Kolb, S. J., & Kissel, J. T. (2015, November). Spinal muscular atrophy. *Neurologic Clinics, 33*(4), 831.

Kramarz, C., Murphy, E., Reilly, M. M., & Rossor, A. M. (2024, January 1). Nutritional peripheral neuropathies. *Journal of Neurology, Neurosurgery & Psychiatry, 95*(1), 61–72.

Krause, K., Wulf, M., Sommer, P., Barkovits, K., Vorgerd, M., Marcus, K., & Eggers, B. (2021, August 24). CSF diagnostics: A potentially valuable tool in neurodegenerative and inflammatory disorders involving motor neurons: A review. *Diagnostics, 11*(9), 1522.

Kubota, A., Ishiura, H., Mitsui, J., Sakuishi, K., Iwata, A., Yamamoto, T., Nishino, I., Tsuji, S., & Shimizu, J. (2018, March 15). A homozygous LAMA2 mutation of c. 818G>A caused partial merosin deficiency in a Japanese patient. *Internal Medicine, 57*(6), 877–882.

Kumar, N., Gross, J. B., Jr., & Ahlskog, J. E. (2004, July 13). Copper deficiency myelopathy produces a clinical picture like subacute combined degeneration. *Neurology, 63*(1), 33–39.

Kwon, B. K., Tetreault, L. A., Martin, A. R., Arnold, P. M., Marco, R. A., Newcombe, V. F., Zipser, C. M., McKenna, S. L., Korupolu, R., Neal, C. J., & Saigal, R. (2024, March). A clinical practice guideline for the management of patients with acute spinal cord injury: Recommendations on hemodynamic management. *Global Spine Journal, 14*(3_suppl), 187S–211S.

Lange, L. M., Gonzalez-Latapi, P., Rajalingam, R., Tijssen, M. A., Ebrahimi-Fakhari, D., Gabbert, C., Ganos, C., Ghosh, R., Kumar, K. R., Lang, A. E., & Rossi, M. (2022, May). Nomenclature of genetic movement disorders: Recommendations of the International Parkinson and Movement Disorder Society task force – An update. *Movement Disorders, 37*(5), 905–935.

Lemmers, R. J., Van der Vliet, P. J., Klooster, R., Sacconi, S., Camaño, P., Dauwerse, J. G., Snider, L., Straasheijm, K. R., Jan van Ommen, G., Padberg, G. W., & Miller, D. G. (2010, September

24). A unifying genetic model for facioscapulohumeral muscular dystrophy. *Science, 329*(5999), 1650–1653.

Leon-Monzon, M. E., & Dalakas, M. C. (1995, May 1). Detection of poliovirus antibodies and poliovirus genome in patients with the post-polio syndrome. *Annals of the New York Academy of Sciences, 753*, 208–218.

Li Hi Shing, S., Chipika, R. H., Finegan, E., Murray, D., Hardiman, O., & Bede, P. (2019, July 16). Post-polio syndrome: More than just a lower motor neuron disease. *Frontiers in Neurology, 10*, 773.

Li Hi Shing, S., Lope, J., Chipika, R. H., Hardiman, O., & Bede, P. (2021, November). Extra-motor manifestations in post-polio syndrome (PPS): Fatigue, cognitive symptoms and radiological features. *Neurological Sciences, 42*(11), 4569–4581.

Li, J., Li, X., Wang, L., & Wu, G. (2022, February 1). Spinal muscular atrophy type IIIb complicated by Moyamoya syndrome: A case report and literature review. *Frontiers in Cellular Neuroscience, 16*, 811596.

Liu, X., Xu, Y., An, M., & Zeng, Q. (2019, February 20). The risk factors for diabetic peripheral neuropathy: A meta-analysis. *PLoS One, 14*(2), e0212574.

Liu, C., Yang, D., Luo, L., Ma, X., Chen, X., Liao, Y., Ning, G., & Qu, H. (2024, October 11). Low bone mineral density and its influencing factors in spinal muscular atrophy without disease-modifying treatment: A single-centre cross-sectional study. *BMC Pediatrics, 24*(1), 651.

Loavenbruck, A. J., Singer, W., Mauermann, M. L., Sandroni, P., B. Dyck, P. J., Gertz, M., Klein, C. J., & Low, P. A. (2016, September). Transthyretin amyloid neuropathy has earlier neural involvement but better prognosis than primary amyloid counterpart: An answer to the paradox? *Annals of Neurology, 80*(3), 401–411.

Long, P. P., Sun, D. W., & Zhang, Z. F. (2022, January). Risk factors for tracheostomy after traumatic cervical spinal cord injury: A 10-year study of 456 patients. *Orthopaedic Surgery, 14*(1), 10–17.

Lorson, C. L., Rindt, H., & Shababi, M. (2010, April 15). Spinal muscular atrophy: Mechanisms and therapeutic strategies. *Human Molecular Genetics, 19*(R1), R111–R118.

Lucci, V. E., Inskip, J. A., McGrath, M. S., Ruiz, I., Lee, R., Kwon, B. K., & Claydon, V. E. (2021, February 1). Longitudinal assessment of autonomic function during the acute phase of spinal cord injury: Use of low-frequency blood pressure variability as a quantitative measure of autonomic function. *Journal of Neurotrauma, 38*(3), 309–321.

Luciano, C. A., Sivakumar, K., Spector, S. A., & Dalakas, M. C. (1995, May 1). Reinnervation in clinically unaffected muscles of patients with Prior paralytic poliomyelitis. *Annals of the New York Academy of Sciences, 753*(1), 394–401.

Malviya, R., & Rajput, S. (2025, October 7). *Artificial intelligence for neural health: Diagnosis and treatment*. CRC Press.

Malviya, R., Rajput, S., & Vaidya, M. (2024, February 21). *Artificial intelligence for bone disorder: Diagnosis and treatment*. Wiley.

Mathis, S., Beauvais, D., Duval, F., Solé, G., & Le Masson, G. (2024, July). The various forms of hereditary motor neuron disorders and their historical descriptions. *Journal of Neurology, 271*(7), 3978–3990.

Mauermann, M. L., & Staff, N. P. (2025). Peripheral neuropathy: A review. *Journal of the American Medical Association.*

McCauley, M. E., O'Rourke, J. G., Yáñez, A., Markman, J. L., Ho, R., Wang, X., Chen, S., Lall, D., Jin, M., Muhammad, A. K., & Bell, S. (2020, September 3). C9orf72 in myeloid cells suppresses STING-induced inflammation. *Nature, 585*(7823), 96–101.

McKenna, M. C., Corcia, P., Couratier, P., Siah, W. F., Pradat, P. F., & Bede, P. (2021, August 16). Frontotemporal pathology in motor neuron disease phenotypes: Insights from neuroimaging. *Frontiers in Neurology, 12*, 723450.

Meisel, A., Saccà, F., Spillane, J., Vissing, J., & MG Collegium Sub-committee. (2024, July). Expert consensus recommendations for improving and standardising the assessment of patients with generalised myasthenia gravis. *European Journal of Neurology, 31*(7), e16280.

Melin, E., Lindroos, E., Lundberg, I. E., Borg, K., & Korotkova, M. (2014). Elevated expression of prostaglandin E2 synthetic pathway in skeletal muscle of prior polio patients. *Journal of Rehabilitation Medicine, 46*(1), 67–72.

Mercuri, E., Signorovitch, J. E., Swallow, E., Song, J., Ward, S. J., Pane, M., Mazzone, E., Messina, S., Vita, G. L., Sormani, M. P., & D'Amico, A. (2016, September 1). Categorizing natural history trajectories of ambulatory function measured by the 6-minute walk distance in patients with Duchenne muscular dystrophy. *Neuromuscular Disorders, 26*(9), 576–583.

Mercuri, E., Bönnemann, C. G., & Muntoni, F. (2019, November 30). Muscular dystrophies. *The Lancet, 394*(10213), 2025–2038.

Mežnarić, S., Belančić, A., Rački, V., Vitezić, D., Mršić-Pelčić, J., & Pilipović, K. (2025, July 22). Prenatal management of spinal muscular atrophy in the era of genetic screening and emerging opportunities in in utero therapy. *Biomedicine, 13*(8), 1796.

Muhamad, R., Akrivaki, A., Papagiannopoulou, G., Zavridis, P., & Zis, P. (2023, January). The role of vitamin B6 in peripheral neuropathy: A systematic review. *Nutrients, 15*(13), 2823.

Muntoni, F., Brockington, M., Blake, D. J., Torelli, S., & Brown, S. C. (2002, November 2). Defective glycosylation in muscular dystrophy. *The Lancet, 360*(9343), 1419–1421.

Murdock, B. J., Zhou, T., Kashlan, S. R., Little, R. J., Goutman, S. A., & Feldman, E. L. (2017, December 1). Correlation of peripheral immunity with rapid amyotrophic lateral sclerosis progression. *JAMA Neurology, 74*(12), 1446–1454.

Murphy, A. P., & Straub, V. (2015). The classification, natural history and treatment of the limb girdle muscular dystrophies. *Journal of Neuromuscular Diseases, 2*, S7–S19.

Murphy, S. M., Laura, M., Fawcett, K., Pandraud, A., Liu, Y. T., Davidson, G. L., Rossor, A. M., Polke, J. M., Castleman, V., Manji, H., & Lunn, M. P. (2012, July 1). Charcot–Marie–Tooth disease: Frequency of genetic subtypes and guidelines for genetic testing. *Journal of Neurology, Neurosurgery & Psychiatry, 83*(7), 706–710.

Nagy, H., & Veerapaneni, K. D. (2023, August 14). Myopathy. In *StatPearls [Internet]*. StatPearls Publishing.

National Spinal Cord Injury Statistical Center. (2019). *The 2019 Annual statistical report for the spinal cord injury model systems*. National Spinal Cord Injury Statistical Center.

Ng, P. S., Dyck, P. J., Laughlin, R. S., Thapa, P., Pinto, M. V., & Dyck, P. J. (2019, March 12). Lumbosacral radiculoplexus neuropathy: Incidence and the association with diabetes mellitus. *Neurology, 92*(11), e1188–e1194.

Nicholson, K., Murphy, A., McDonnell, E., Shapiro, J., Simpson, E., Glass, J., Mitsumoto, H., Forshew, D., Miller, R., & Atassi, N. (2018, January). Improving symptom management for people with amyotrophic lateral sclerosis. *Muscle & Nerve, 57*(1), 20–24.

Nishio, H., Niba, E. T., Saito, T., Okamoto, K., Takeshima, Y., & Awano, H. (2023, July 26). Spinal muscular atrophy: The past, present, and future of diagnosis and treatment. *International Journal of Molecular Sciences, 24*(15), 11939.

Ofori, E., Ramai, D., Ona, M., & Reddy, M. (2017, August 31). Paraneoplastic dermatomyositis syndrome presenting as dysphagia. *Gastroenterology Research, 10*(4), 251.

On, A. Y., Latifoglou, E., Çınar, E., & Tanıgör, G. (2024, May 1). Prevalence and severity of central sensitization in post-polio syndrome: Associations with clinical measures and quality of life. *Annals of Indian Academy of Neurology, 27*(3), 282–288.

Onders, R. P., Khansarinia, S., Ingvarsson, P. E., Road, J., Yee, J., Dunkin, B., & Ignagni, A. R. (2022, October). Diaphragm pacing in spinal cord injury can significantly decrease mechanical ventilation in multicenter prospective evaluation. *Artificial Organs, 46*(10), 1980–1987.

Oude Lansink, I. L., Gorter, J. W., van der Pol, W. L., Scheiberlich, P. J., van Eijk, R. P., Bartels, B., & Beelen, A. (2025, December). Course of joint range of motion in children with spinal muscular atrophy receiving disease-modifying treatment. *Orphanet Journal of Rare Diseases, 20*(1), 592.

Panagiotou, P., Kanaka-Gantenbein, C., & Kaditis, A. G. (2022, August 11). Changes in ventilatory support requirements of spinal muscular atrophy (SMA) patients post gene-based therapies. *Children, 9*(8), 1207.

Paul, G. R., Gushue, C., Kotha, K., & Shell, R. (2021, April). The respiratory impact of novel therapies for spinal muscular atrophy. *Pediatric Pulmonology, 56*(4), 721–728.

Pender, N., Pinto-Grau, M., & Hardiman, O. (2020, October 1). Cognitive and behavioural impairment in amyotrophic lateral sclerosis. *Current Opinion in Neurology, 33*(5), 649–654.

Pereira, A. S., Gatti, M., Ribeiro, V. V., Taveira, K. V., & Berretin-Felix, G. (2024). Speech language pathology interventions in the areas of breathing, chewing, swallowing and speaking: A scoping review. *CoDAS, 36*, e20220339.

Plomp, L., Chassepot, H., Psimaras, D., Maisonobe, T., Mensi, E., Leonard-Louis, S., Plu, I., Rozes, A., Tubach, F., Touat, M., & Anquetil, C. (2025, March 1). Features of myositis and myasthenia gravis in patients treated with immune checkpoint inhibitors: A multicentric, retrospective cohort study. *The Lancet Regional Health–Europe, 50*.

Pokrishevsky, E., Grad, L. I., & Cashman, N. R. (2016, March 1). TDP-43 or FUS-induced misfolded human wild-type SOD1 can propagate intercellularly in a prion-like fashion. *Scientific Reports, 6*(1), 22155.

Querin, G., El Mendili, M. M., Lenglet, T., Behin, A., Stojkovic, T., Salachas, F., Devos, D., Le Forestier, N., del Mar Amador, M., Debs, R., & Lacomblez, L. (2019, January 1). The spinal and cerebral profile of adult spinal-muscular atrophy: A multimodal imaging study. *NeuroImage: Clinical, 21*, 101618.

Raab, A. M., Mueller, G., Elsig, S., Gandevia, S. C., Zwahlen, M., Hopman, M. T., & Hilfiker, R. (2021, December 31). Systematic review of incidence studies of pneumonia in persons with spinal cord injury. *Journal of Clinical Medicine, 11*(1), 211.

Radha, G., & Lopus, M. (2021, September 1). The spontaneous remission of cancer: Current insights and therapeutic significance. *Translational Oncology, 14*(9), 101166.

Rajput, S., Malviya, R., Bahadur, S., & Puri, D. (2023, October 1). Recent updates on the development of therapeutics for the targeted treatment of Alzheimer's disease. *Current Pharmaceutical Design, 29*(35), 2802–2813.

Ramachandran, A. K., Goodman, S. P., Jackson, M. J., & Lathlean, T. J. (2021, April 19). Effects of muscle strengthening and cardiovascular fitness activities for poliomyelitis survivors: A systematic review and meta-analysis. *Journal of Rehabilitation Medicine, 53*(4), 2789.

Raman, S. V., Hor, K. N., Mazur, W., Halnon, N. J., Kissel, J. T., He, X., Tran, T., Smart, S., McCarthy, B., Taylor, M. D., & Jefferies, J. L. (2015, February 1). Eplerenone for early cardiomyopathy in Duchenne muscular dystrophy: A randomised, double-blind, placebo-controlled trial. *The Lancet Neurology, 14*(2), 153–161.

Rana, A., Malviya, R., Rajput, S., Sridhar, S. B., & Wadhwa, T. (2025). Trends in nanoparticle-based strategies for the management of neuroinflammation. *CNS and Neurological Disorders - Drug Targets*.

Record, C. J., Pipis, M., Skorupinska, M., Blake, J., Poh, R., Polke, J. M., Eggleton, K., Nanji, T., Zuchner, S., Cortese, A., & Houlden, H. (2024, September). Whole genome sequencing increases the diagnostic rate in Charcot-Marie-Tooth disease. *Brain, 147*(9), 3144–3156.

Remijn-Nelissen, L., Verschuuren, J. J., & Tannemaat, M. R. (2022, October 1). The effectiveness and side effects of pyridostigmine in the treatment of myasthenia gravis: A cross-sectional study. *Neuromuscular Disorders, 32*(10), 790–799.

Ricotti, V., Mandy, W. P., Scoto, M., Pane, M., Deconinck, N., Messina, S., Mercuri, E., Skuse, D. H., & Muntoni, F. (2016, January). Neurodevelopmental, emotional, and behavioural problems in Duchenne muscular dystrophy in relation to underlying dystrophin gene mutations. *Developmental Medicine & Child Neurology, 58*(1), 77–84.

Ringholz, G. M., Appel, S. H., Bradshaw, M., Cooke, N. A., Mosnik, D. M., & Schulz, P. E. (2005, August 23). Prevalence and patterns of cognitive impairment in sporadic ALS. *Neurology, 65*(4), 586–590.

Rodrigues, L. C., & Lockwood, D. N. (2011, June 1). Leprosy now: Epidemiology, progress, challenges, and research gaps. *The Lancet Infectious Diseases, 11*(6), 464–470.

Rosenbohm, A., Liu, M., Nagel, G., Peter, R. S., Cui, B., Li, X., Kassubek, J., Rothenbacher, D., Lulé, D., Cui, L., & Ludolph, A. C. (2018, April). Phenotypic differences of amyotrophic lateral sclerosis (ALS) in China and Germany. *Journal of Neurology, 265*(4), 774–782.

Rosenthal, L. S. (2022, October 1). Neurodegenerative cerebellar ataxia. *Continuum: Lifelong Learning in Neurology, 28*(5), 1409–1434.

Russo, T., & Riessland, M. (2022, June 2). Age-related midbrain inflammation and senescence in Parkinson's disease. *Frontiers in Aging Neuroscience, 14*, 917797.

Saldert, C., Ferm, U., & Hartelius, L. (2021, September 23). Huntington's disease. In *Handbook of pragmatic language disorders: Complex and underserved populations* (pp. 461–494). Springer.

Sandberg, A. (2023, January 1). Motor unit properties do not correlate between MUNIX and needle EMG in remote polio in the biceps brachii muscle. *Clinical Neurophysiology Practice, 8*, 24–31.

Sanjari, E., Raeisi Shahraki, H. G., Khachatryan, L., & Mohammadian-Hafshejani, A. (2024, April 16). Investigating the association between diabetes and carpal tunnel syndrome: A systematic review and meta-analysis approach. *PLos One, 19*(4), e0299442.

Savarese, M., Di Fruscio, G., Tasca, G., Ruggiero, L., Janssens, S., De Bleecker, J., Delpech, M., Musumeci, O., Toscano, A., Angelini, C., & Sacconi, S. (2015, July 1). Next generation sequencing on patients with LGMD and nonspecific myopathies: Findings associated with ANO5 mutations. *Neuromuscular Disorders, 25*(7), 533–541.

Schindler, R. F., Scotton, C., Zhang, J., Passarelli, C., Ortiz-Bonnin, B., Simrick, S., Schwerte, T., Poon, K. L., Fang, M., Rinne, S., & Froese, A. (2016, January 4). POPDC1 S201F causes muscular dystrophy and arrhythmia by affecting protein trafficking. *The Journal of Clinical Investigation, 126*(1), 239–253.

Sharma, A., Yu, C., Leung, C., Trane, A., Lau, M., Utokaparch, S., Shaheen, F., Sheibani, N., & Bernatchez, P. (2010, November 1). A new role for the muscle repair protein dysferlin in endothelial cell adhesion and angiogenesis. *Arteriosclerosis, Thrombosis, and Vascular Biology, 30*(11), 2196–2204.

Shuey, N. H. (2022, February 17). Ocular myasthenia gravis: A review and practical guide for clinicians. *Clinical and Experimental Optometry, 105*(2), 205–213.

Singh, S. (2020, April 1). Respiratory symptoms and signs. *Medicine, 48*(4), 225–233.

Skre, H. (1974, August). Genetic and clinical aspects of Charcot-Marie-Tooth's disease. *Clinical Genetics, 6*(2), 98–118.

Stahl, J. H., Konle, M., Kowarik, M., Dubois, E., Armbruster, M., Kleiser, B., Grimm, A., Martin, P., & Marquetand, J. (2025, May 15). Prevalence and course of muscle-specific receptor tyrosine kinase (MuSK) antibodies in myasthenia gravis - A retrospective study. *Journal of the Neurological Sciences, 472*, 123449.

Stankovic, I., Fanciulli, A., Sidoroff, V., & Wenning, G. K. (2023, October). A review on the clinical diagnosis of multiple system atrophy. *The Cerebellum, 22*(5), 825–839.

Straub, V., Murphy, A., Udd, B., Corrado, A., Aymé, S., Bönneman, C., de Visser, M., Hamosh, A., Jacobs, L., Khizanishvili, N., & Kroneman, M. (2018, August 1). 229th ENMC international workshop: Limb girdle muscular dystrophies–nomenclature and reformed classification Naarden, The Netherlands, 17–19 March 2017. *Neuromuscular Disorders, 28*(8), 702–710.

Sugarman, E. A., Nagan, N., Zhu, H., Akmaev, V. R., Zhou, Z., Rohlfs, E. M., Flynn, K., Hendrickson, B. C., Scholl, T., Sirko-Osadsa, D. A., & Allitto, B. A. (2012, January). Panethnic carrier screening and prenatal diagnosis for spinal muscular atrophy: Clinical laboratory analysis of >72 400 specimens. *European Journal of Human Genetics, 20*(1), 27–32.

Szczęsny, P., Świerkocka, K., & Olesińska, M. (2018, October 31). Differential diagnosis of idiopathic inflammatory myopathies in adults–the first step when approaching a patient with muscle weakness. *Reumatologia/Rheumatology, 56*(5), 307–315.

Taniguchi-Ikeda, M., Morioka, I., Iijima, K., & Toda, T. (2016, October 1). Mechanistic aspects of the formation of α-dystroglycan and therapeutic research for the treatment of α-dystroglycanopathy: A review. *Molecular Aspects of Medicine, 51*, 115–124.

Tanrıverdi, M., Düzenli, S., Gündüz, M. S., Arslantürk, A., & Erekdağ, A. (2025, January 1). Investigation of neurogenic dysphagia in commonly seen neurological diseases. *Anatolian Clinic the Journal of Medical Sciences, 30*(1), 150–161.

Taylor, S. W., Laughlin, R. S., Kumar, N., Goodman, B., Klein, C. J., Dyck, P. J., & Dyck, P. J. (2017, October 1). Clinical, physiological and pathological characterisation of the sensory predominant peripheral neuropathy in copper deficiency. *Journal of Neurology, Neurosurgery & Psychiatry, 88*(10), 839–845.

Thanvi, B. R., & Lo, T. C. (2004, December). Update on myasthenia gravis. *Postgraduate Medical Journal, 80*(950), 690–700.

Toniolo, A., Genoni, A., Maccari, G., Chumakov, K., Basolo, F., Bono, G., Mauri, M., Molteni, F., Arrondini, L., Bertolasi, L., & Monaco, S. (2025, September). Low-grade persistent poliovirus infection in long-term polio survivors diagnosed with post-polio syndrome: Diagnostic and clinical implications. *Journal of Neurology, 272*(9), 1–7.

Trabacca, A., Ferrante, C., Oliva, M. C., Fanizza, I., Gallo, I., & De Rinaldis, M. (2024, October 21). Update on inherited pediatric motor neuron diseases: Clinical features and outcome. *Genes, 15*(10), 1346.

Tracy, J. A. (2023, October 1). Autoimmune axonal neuropathies. *Continuum: Lifelong Learning in Neurology, 29*(5), 1378–1400.

Tsai, S. R., & Lin, W. H. (2025, April 1). Acute cauda Equina syndrome following spinal anesthesia during cesarean section: A case report. *Acta Neurologica Taiwanica, 34*(2), 98–101.

Urciuolo, A., Quarta, M., Morbidoni, V., Gattazzo, F., Molon, S., Grumati, P., Montemurro, F., Tedesco, F. S., Blaauw, B., Cossu, G., & Vozzi, G. (2013, June 7). Collagen VI regulates satellite cell self-renewal and muscle regeneration. *Nature Communications, 4*(1), 1964.

Van Der Heul, A. M., Van Eijk, R. P., Wadman, R. I., Asselman, F., Cuppen, I., Nievelstein, R. A., Gerrits, E., van der Pol, W. L., & Van Den Engel-Hoek, L. (2022, August). Mastication in patients with spinal muscular atrophy types 2 and 3 is characterized by abnormal efficiency, reduced endurance, and fatigue. *Dysphagia, 37*(4), 715–723.

Vignesh, G., Balachandran, K., Kamalanathan, S., & Hamide, A. (2013, March 1). Myoedema: A clinical pointer to hypothyroid myopathy. *Indian Journal of Endocrinology and Metabolism, 17*(2), 352.

Vitale, M., Roye, B., Bloom, Z., Kunes, J. A., Matsumoto, H., Roye, D., Farrington, D., Flynn, J., Halanski, M., Hasler, C., & Miladi, L. (2022, February 1). Best practices for the orthopaedic care of children with spinal muscular atrophy: A consensus statement from the European neuromuscular Centre standard of care orthopaedic working group. *Journal of the Pediatric Orthopaedic Society of North America, 4*(1).

Voet, N. B., van der Kooi, E. L., van Engelen, B. G., & Geurts, A. C. (2019). Strength training and aerobic exercise training for muscle disease. *Cochrane Database of Systematic Reviews, 12*(12), CD003907.

Walhout, R., Verstraete, E., Van Den Heuvel, M. P., Veldink, J. H., & Van Den Berg, L. H. (2018, January 2). Patterns of symptom development in patients with motor neuron disease. *Amyotrophic Lateral Sclerosis and Frontotemporal Degeneration, 19*(1–2), 21–28.

Wallace, L. M., Liu, J., Domire, J. S., Garwick-Coppens, S. E., Guckes, S. M., Mendell, J. R., Flanigan, K. M., & Harper, S. Q. (2012, July 1). RNA interference inhibits DUX4-induced muscle toxicity in vivo: Implications for a targeted FSHD therapy. *Molecular Therapy, 20*(7), 1417–1423.

Wang, H., Li, L., Qin, L. L., Song, Y., Vidal-Alaball, J., & Liu, T. H. (2018). Oral vitamin B 12 versus intramuscular vitamin B 12 for vitamin B 12 deficiency. *Cochrane Database of Systematic Reviews, 3*.

Wang, T. Y., Park, C., Zhang, H., Rahimpour, S., Murphy, K. R., Goodwin, C. R., Karikari, I. O., Than, K. D., Shaffrey, C. I., Foster, N., & Abd-El-Barr, M. M. (2021, December 13). Management of acute traumatic spinal cord injury: A review of the literature. *Frontiers in Surgery, 8*, 698736.

Watson, J. C., & Dyck, P. J. (2015, July 1). Peripheral neuropathy: A practical approach to diagnosis and symptom management. *Mayo Clinic Proceedings, 90*(7), 940–951.

Wecht, J. M. (2022, February 1). Management of blood pressure disorders in individuals with spinal cord injury. *Current Opinion in Pharmacology, 62*, 60–63.

Westeneng, H. J., Debray, T. P., Visser, A. E., van Eijk, R. P., Rooney, J. P., Calvo, A., Martin, S., McDermott, C. J., Thompson, A. G., Pinto, S., & Kobeleva, X. (2018, May 1). Prognosis for patients with amyotrophic lateral sclerosis: Development and validation of a personalised prediction model. *The Lancet Neurology, 17*(5), 423–433.

White, N. H., & Black, N. H. (2017). *Spinal cord injury facts and figures at a glance. National Spinal Cord Injury Statistical Center, facts and figures at a glance.* University of Alabama at Birmingham.

Wiechers, D. O., & Hubbell, S. L. (1981, November). Late changes in the motor unit after acute poliomyelitis. *Muscle & Nerve: Official Journal of the American Association of Electrodiagnostic Medicine, 4*(6), 524–528.

Wilson, G. N., Naga, O., & Tonk, V. S. (2025, January 19). Genetic disorders. In *Pediatric board study guide: A last minute review* (pp. 107–146). Springer Nature Switzerland.

Winstock, A. R., & Ferris, J. A. (2020, February). Nitrous oxide causes peripheral neuropathy in a dose dependent manner among recreational users. *Journal of Psychopharmacology, 34*(2), 229–236.

Wong, C., Gregory, J. M., Liao, J., Egan, K., Vesterinen, H. M., Khan, A. A., Anwar, M., Beagan, C., Brown, F. S., Cafferkey, J., & Cardinali, A. (2023, February 1). Systematic, comprehensive, evidence-based approach to identify neuroprotective interventions for motor neuron disease: Using systematic reviews to inform expert consensus. *BMJ Open, 13*(2), e064169.

Xiang, X. N., Zong, H. Y., Ou, Y., Yu, X., Cheng, H., Du, C. P., & He, H. C. (2021, May 24). Exoskeleton-assisted walking improves pulmonary function and walking parameters among individuals with spinal cord injury: A randomized controlled pilot study. *Journal of Neuroengineering and Rehabilitation, 18*(1), 86.

Yellepeddi, V. K., Race, J. A., Mcfarland, M. M., Constance, J. E., Fanaeian, E., & Murphy, N. A. (2024, June 1). Effectiveness of atropine in managing sialorrhea: A systematic review and meta-analysis. *International Journal of Clinical Pharmacology and Therapeutics, 62*(6), 267.

Yonekawa, T., & Nishino, I. (2015, March 1). Ullrich congenital muscular dystrophy: Clinicopathological features, natural history and pathomechanism(s). *Journal of Neurology, Neurosurgery & Psychiatry, 86*(3), 280–287.

Yoskovitz, G., Peled, Y., Gramlich, M., Lahat, H., Resnik-Wolf, H., Feinberg, M. S., Afek, A., Pras, E., Arad, M., Gerull, B., & Freimark, D. (2012, June 1). A novel titin mutation in adult-onset familial dilated cardiomyopathy. *The American Journal of Cardiology, 109*(11), 1644–1650.

Zakharova, M. (2021, September). Modern approaches in gene therapy of motor neuron diseases. *Medicinal Research Reviews, 41*(5), 2634–2655.

Zambon, A. A., Pini, V., Bosco, L., Falzone, Y. M., Munot, P., Muntoni, F., & Previtali, S. C. (2023, March 1). Early onset hereditary neuronopathies: An update on non-5q motor neuron diseases. *Brain, 146*(3), 806–822.

Zappa, G., LoMauro, A., Baranello, G., Cavallo, E., Corti, P., Mastella, C., & Costantino, M. A. (2021, December). Intellectual abilities, language comprehension, speech, and motor function in children with spinal muscular atrophy type 1. *Journal of Neurodevelopmental Disorders, 13*(1), 9.

Zhang, Y., Al Mamun, A., Yuan, Y., Lu, Q., Xiong, J., Yang, S., Wu, C., Wu, Y., & Wang, J. (2021, June). Acute spinal cord injury: Pathophysiology and pharmacological intervention. *Molecular Medicine Reports, 23*(6), 417.

Chapter 5
Sensory Organ Disorders and Associated Disabilities

Abstract The sense organs provide an input and response to the environment by sensing the external stimuli and sending the information to the brain. This chapter examines sensory organ abnormalities and sensory organ disabilities. It analyses the consequence of being deficient in vision, hearing, taste, smell, and touch in communication, learning, and daily activities. The chapter explores the cause and the effect of sensory disability, as well as the importance of timely detection and management. In this chapter, it is also described that people who have a disorder of the sense organs, like visual impairment, hearing problems, and auditory processing disorders, generally have social difficulties such as communication barriers, lack of adequate information and mobility, social isolation, and lack of involvement in the community life of a person and the family.

Keywords Sensory organ disorders · Sensory disability · Visual impairment · Hearing loss · Auditory processing disorders · Tactile impairment · Communication barriers

5.1 Introduction

The malfunctions of the sensory organs and related impairments are key to human functioning because the sensory organs are the main connecting point through which people sense, interpret, and experience the environment. All communication, learning, movement, safety, emotional experience, and even social interaction are possible with the help of vision, hearing, touch, taste, and smell. These sensory systems are interrupted, leading to distortions of the channels of sensory information to the brain, not only causing physiological disruption but also functional limitation, which has ramifications on everyday activity and social engagement (Koziol et al., 2011). The disorders of the sensory organs are not isolated medical conditions; rather, they are a complex experience of the disability, being affected by biological, psychological, and environmental factors. Sensory impairment is recognized

R. Malviya, S. Rajput, *Neurogenetic and Neurodevelopmental Disabilities*, SpringerBriefs in Modern Perspectives on Disability Research,
https://doi.org/10.1007/978-981-95-9223-4_5

as one of the major causes of disability burden in the world, according to modern disability studies. Disability in modern models, such as in the International Classification of Functioning, Disability and Health provided by the World Health Organization, is not only the consequence of the sensory loss but also a deriving interaction between impairment, restriction of activity, restriction of participation, and the limitation of the context (World Health Organization, 2007). The framework can be especially relevant when examining sensory disorders, as individuals with comparable levels of sensory impairment may experience vastly different degrees of disability. This variation often depends on their access to assistive technology, rehabilitation services, opportunities for inclusion, and the presence of supportive social attitudes. The disabled sensory functioning is impairing because environmental and societal circumstances do not support the modified sensory capacities (Carvill, 2001).

The contribution of disorders to the sensory organs is huge, with senses being the foundations of almost every aspect of human existence. Language development, educational attainment, employment, and social communication all require vision and hearing. Tactile imparts the ability to coordinate motor activities, handle objects, and feel safe, whereas the gustatory and olfactory senses are critical in nutrition, appetite management, safety consciousness, and feeling pleasant (Piotrowska et al., 2024). In case of disruption of one or more of these sensory systems, people can find it difficult to carry out their activities of daily living, to access information, to navigate physical environments, to sustain interpersonal interactions, and to engage in an economic and cultural life in their entirety. Usually, these limitations affect not only the individual but also their families, caregivers, and communities. Throughout the lifespan, sensory disabilities can progress in various ways. Congenital sensory impairments may be genetic or acquired as a result of ageing, chronic disease, infection, trauma, or environmental exposure and are acquired by some individuals or others who are born with the condition (Shave et al., 2022). The slow loss of sensory functions that can accompany old age (such as deafness or hearing loss), diabetes (retinopathy), or peripheral neuropathy (or other conditions) may result in a situation where the disability develops without being detected and treated early enough (Völter et al., 2021). In comparison, functional capacity, psychological distress, and premature loss of autonomy can result from acute loss of sensory input due to trauma, stroke, or acute infection. The two trends indicate the significance of early detection, preventive measures, and rehabilitation processes.

Sensory impairments are often accompanied by other health issues, which enhance disability and also make the care requirements more difficult (Sterkenburg et al., 2022). There is a high correlation between visual impairment and risk of falls, fractures, depression, and cognitive impairment, especially in the elderly (Dinarvand et al., 2024). Social isolation, communication breakdown, diminished occupational performance, and faster deterioration of cognition have been associated with hearing impairment. The inability to feel, particularly in metabolic and neurological conditions, predisposes to burns, wounds, infections, and amputation of the limbs. Taste and smell impairments can lead to nutritional impairment, loss of appetite, and the inability to sense hazards in the environment like rotten food or gas leaks (Fjaeldstad & Smith, 2022). These associated effects demonstrate that sensory-related disability can be multidimensional in nature and cumulative in nature. A sensory disability can entail the lack of one or several of the senses, such as sight, hearing, touch, taste, or smell experience, and sensory perception and processing is distorted (Fabio et al., 2024). Sensory disabilities are not necessarily observable, and several individuals live with sensory disabilities without displaying any outward expression to draw the attention of others to their predicament. The levels of disability are diverse, in the form of mild impaired functional limit to severe reliance on external assistance. People might need to make great adjustments to engage in schooling, work, and community life, and others work on their own with slight changes. The majority of individuals with sensory impairments lie within this continuum with varying needs based on situations and surroundings.

The difference between sensory impairment and sensory disability is crucial to the explanation of lived experience. Sensory impairment refers to a partial or complete loss of one or more senses, such as vision, hearing, tactile sensation, or alterations in taste and smell (Tam & Koppel, 2021). On the other hand, sensory impairment manifests when it hinders daily activities, social interactions, and opportunities (Boldsen, 2022). An example is that visual impairment may be complete or partial blindness or loss of vision, both in the recognition of details and the perception of colour. The hearing loss can be mild hearing loss, moderate hearing loss, or severe deafness, which impacts the perception of speech and communication. Impairments that affect the sense of touch, temperature, and pain are known as tactile impairments; the impairments that affect the sense of flavour and odour recognition are known as gustatory and olfactory impairments (Trujillo Tanner et al., 2022). Each of these impairments has different functional implications, yet they are all likely to cause disability in the case of the lack of support. Individuals who experience sensory impairments tend to have problems with articulation, movement, and accessibility of information. Hearing and vision impairments could limit verbal and non-verbal communication and require alternative communication such as sign language, Braille, audio, or assistive technology. The reduced sensory output can lead to the loss of mobility and orientation and the increased dependence on training and adaptive gadgets or environmental support (Haanes, 2023). Access to information is particularly difficult where there is no availability of education resources, internet-based information, and governmental resources that are not available in the sensory mode. The problems also demonstrate the role of inclusive design, universal accessibility, and supportive services in reducing disability.

Sensory disability diagnosis is complex and heterogeneous. They may be inborn as a result of mutations of the genes or prenatal exposure or acquired as a result of infections, chronic diseases, metabolic diseases, aging, environmental toxins, or physical injuries (Alotaibi et al., 2025). Even though they are diverse in their causes, the sensory disabilities are prevalent through a shared neurological mechanism that proves the abnormality of transmitting or processing sensory signals. Diagnosis is generally a clinical analysis by specifically trained experts such as ophthalmologists, audiologists, neurologists, or rehabilitation specialists, who are capable of making a judgment not only of the functionality of the senses but also in terms of their impact on their daily existence (DeBonis & Donohue, 2024). The management and care of sensory disabilities extend beyond medical care and are more about rehabilitation, adaptation, and empowerment. While certain levels of sensory impairment can be treated or even cured with medical or surgical procedures, most cases require a long-term approach to address them. This may involve the use of assistive devices, sensory training, environmental adjustments, and psychosocial support (Section on Complementary and Integrative Medicine, Council on Children with Disabilities, et al., 2012). Rehabilitation services are significant in ensuring independence, the restoration of functional capacities, and quality living. Ensuring equity, social inclusivity, and access is crucial to prevent unnecessary barriers from limiting individuals with sensory impairments. To develop effective healthcare policies, education, workplace accommodations, and social support, it is essential to view sensory organ disorders within the broader context of disability. There is a necessity in the treatment of sensory disability with the combination of clinical care and rehabilitation, providing assistive technology, inclusion, and attitudinal change. This chapter emphasizes comprehensive strategies that provide complete support, honour, and ensure the well-being of individuals with sensory disabilities across their lifespan while recognizing the multifaceted and lived experiences of sensory impairment. Tables 5.1 and 5.2 provide illustrations of various sensory organ disorders and disabilities.

5.2 Vision Impairment and Disability

One of the most important sensory disorders is vision impairment, as it influences the functioning of a human and has much further-reaching consequences than the loss of vision (Petty, 2021). Disability-wise, the vision impairment cannot be regarded as the reduction of the visual acuity or losing sight, but the level of interference of visual loss disrupts the everyday functioning of the individual and full-scale participation in society. The World Health Organization defines disability in the International Classification of Functioning, Disability, and Health (ICF) as the interaction between an individual's impairment and environmental, social, and attitudinal barriers (World Health Organization, 2007). In the case of vision loss, this interaction will draw the distinction between the minor inconvenience and the extreme disability that an individual will go through. Vision impairment is a continuum that ranges from low vision to complete blindness, and it is characterized by varying degrees of functional impairment (Rajput et al., 2024). Albeit some of them still have some residual vision, they can still face considerable problems performing tasks that demand accuracy, contrast sensitivity, depth perception, or peripheral awareness. Researchers witness cases of disability where the resulting visual impairments restrict autonomy, productivity, or aggressive attention. Even people with equal degrees of visual impairment may have extremely different disability outcomes in terms of access to rehabilitation services, assistive technology, inclusive environments, and social attitudes toward disability. The resulting disability due to the impairment of vision is mostly progressive and accumulative

Table 5.1 Sensory organ disorders and associated disabilities

Sensory impairment	Primary sensory deficit	Major associated disabilities	Functional challenges	Psychosocial and participation challenges	Long-term consequences
Vision impairment/ blindness	Reduced or absent visual perception	Functional disability, mobility disability, educational disability, occupational disability	Difficulty in reading, navigation, face recognition, self-care, safe mobility, and environmental awareness	Social isolation, dependency, stigma, reduced employment opportunities, depression and anxiety	Reduced quality of life, increased fall risk, economic dependency, chronic mental health disorders
Hearing impairment/ deafness	Partial or complete loss of auditory perception	Communication disability, educational disability, social participation restriction	Difficulty understanding speech, especially in noisy environments; impaired language development; reduced auditory awareness	Social withdrawal, communication barriers, marginalization, identity challenges, emotional distress	Delayed cognitive and language development (early onset), unemployment, social exclusion
Tactile (touch) impairment	Reduced or absent sensation of touch, pressure, pain, temperature	Sensory-motor disability, safety-related disability	Impaired object manipulation, difficulty detecting injuries, burns, pressure sores, reduced fine motor control	Increased dependence, fear of injury, reduced confidence in physical interaction	Chronic wounds, secondary infections, long-term mobility limitations
Gustatory (taste) impairment	Partial or complete loss of taste perception	Nutritional disability, health management disability	Reduced appetite, inability to detect spoiled food, difficulty maintaining balanced nutrition	Reduced enjoyment of eating, social withdrawal from shared meals, emotional dissatisfaction	Malnutrition, weight loss or gain, worsening of chronic illnesses
Olfactory (smell) impairment	Reduced or absent sense of smell	Safety disability, nutritional disability	Inability to detect smoke, gas leaks, spoiled food, or environmental hazards	Anxiety related to safety, reduced food enjoyment, social insecurity	Increased accident risk, nutritional deficiencies, reduced life satisfaction
Combined sensory impairment (e.g. deafblindness)	Simultaneous impairment of two or more senses	Severe functional disability, communication disability, total participation restriction	Extreme difficulty in communication, orientation, mobility, and independent living	Profound isolation, reliance on caregivers, limited social interaction	Lifelong dependency, restricted educational and occupational outcomes

(continued)

Table 5.1 (continued)

Sensory impairment	Primary sensory deficit	Major associated disabilities	Functional challenges	Psychosocial and participation challenges	Long-term consequences
Early-onset sensory impairment	Sensory loss during childhood	Developmental disability, learning disability	Delayed motor, cognitive, and language development	Limited peer interaction, educational exclusion	Lifelong functional and social limitations
Late-onset sensory impairment	Sensory loss during adulthood or aging	Adjustment-related disability, psychological disability	Difficulty adapting to daily activities and technologies	Identity loss, depression, anxiety, reduced autonomy	Accelerated functional decline, mental health deterioration
Progressive sensory disorders	Gradual sensory deterioration	Chronic disability	Increasing difficulty in adaptation and planning	Fear of future dependence, emotional distress	Escalating disability, cumulative functional loss
Sensory impairment with comorbid conditions	Sensory loss with chronic illness	Complex disability	Increased healthcare needs, reduced self-management ability	Caregiver burden, reduced independence	Higher morbidity, increased healthcare costs

Table 5.2 Comparative overview of sensory impairments and disabilities

Sensory system	Key disability domain	Major functional limitation	Primary rehabilitation focus
Vision	Mobility and independence	Navigation, reading	Vision rehabilitation, assistive tech
Hearing	Communication	Speech perception	Auditory rehabilitation
Touch	Motor and safety	Object handling, injury prevention	Sensory integration therapy
Taste	Nutrition	Appetite regulation	Dietary and nutritional support
Smell	Safety and quality of life	Hazard detection	Environmental adaptations

(Marques et al., 2022). Most people can adapt, but they struggle more as they age or go blind. This development renders it relevant to envision vision impairment as a disability that is not a medical problem but rather a lifelong burden altering as time goes on.

5.2.1 Functional Disability

Loss of vision causes functional disability, which means the inability to support everyday living. Sight is most critical in personal hygiene, domestic affairs, and personal security. Even the simple things like dressing, grooming, eating, or taking medication are complicated and prone to errors when visual input has been impaired. People with vision loss can have difficulty differentiating colours, recognizing items, reading signs, and recognizing risks, resulting in a higher reliance on assistance and help (Gamage et al., 2023). The one that is of great concern is reading disability because reading is a basic part of education, employment, health management, and social involvement. Weak visual perception, contrast sensitivity, or visual field impairments impair the capacity to read printed or digital text and have the consequence of information restriction. The restriction frequently results in secondary disability, including lack of health literacy, financial management, and news, legal documents, and social communication (Xu et al., 2024a). Without easily available formats like Braille, large print, and audio texts, people with visual impairment are left out of the information-driven societies. The lack of control over time, finances, and personal affairs makes the situation even worse with the loss of independence. The inability to identify money, read bills, and use digital banking solutions may lead to financial instability and manipulation (Kameswaran & Marathe, 2023). Prolonged dependency on others due to the necessity to get basic needs may cause low self-esteem, low autonomy, and a mental addiction, which strengthens the paralysing effect of the vision loss.

5.2.2 Mobility and Environmental Barriers

The most disabling effects of vision impairment are mobility and orientation disabilities. Vision is required to convey the necessary spatial data on the way to move safely and efficiently around the environment. Poor eye vision or blindness in its entirety poses serious problems for individuals estimating distance, identifying obstacles, noticing changes in surfaces, and more or less understanding the surrounding environment (Gao et al., 2022). These complications precondition a high level of risk of falls, collisions, and injuries, particularly when it comes to older adults. Odd environments with no visual aids massively restrict free movement. The necessity to cross the roads, to use the means of transportation, or to walk in the closest area means greater cognitive loads and external assistance in the majority of cases. Consequently, individuals with vision impairments find it challenging to navigate outside of familiar paths, which can result in health issues and a tendency toward social withdrawal. A constraint is one of the factors that result in secondary illness, which includes reduced cardiovascular aerobic capacity, musculoskeletal debilitation, and dependence (Juopperi et al., 2021). Barriers to the environmental factor are critical in making vision impairment a disability. The inadequate infrastructure, the deficiency of tactile paving, insufficient lighting, the deficiency of audible signs, and unreachable spaces of the population aggravate mobility restrictions (Kapsalis et al., 2024). People with residual vision can even develop a viable disability in environments that do not accommodate visual diversity. On the other hand, disability can also be reduced to a considerable extent with the help of inclusive design and accessible environments, which promote the role of society in combating vision-related impairments.

5.2.3 Educational and Occupational Disability

The implications of vision impairment for education and employment are far-reaching and, in many cases, lead to lifelong socioeconomic disadvantage. Vision is one of the main media of learning in the classroom setting, with a significant share of information being obtained by vision. Children with vision issues experience limitations when using textbooks, visual instructional materials, and digital and classroom assignments (Van der Meulen et al., 2022). The high risk of these children achieving poor results at school and dropping out of school is high without proper accommodations, which include Braille teaching, assistive technologies, or trained instructors. Disability in education, in most cases, spills over to adulthood, where higher education and skill acquisition are restricted. This educational marginalization is translated in the form of occupational disability because people with visual impairment face limited access to jobs and are discriminated against in their workplaces (Benavot et al., 2022). Most of the professions are visually stimulating, and the absence of reasonable accommodations, including screen readers, adaptive software, or an accessible workplace, leads to underemployment or unemployment. Career development is usually hindered by the misconception created in society regarding productivity and capability, even when people have the right qualifications. The direct effect of these occupational and educational barriers is economic disability. Loss of income, unemployment, and dependence on social welfare are factors that risk the impoverished individuals with a vision impairment (Marques et al., 2022). It is not only the individual but also the family and the society that bear the economic burden, which is the costs of health care, costs of caregiving, and loss of productivity. Such difficulties in the economy support the cycle of disability, restricting access to healthcare, rehabilitation, and other assistive technologies that otherwise might enhance functional outcomes.

5.2.4 Psychosocial and Developmental Disability

The psychosocial consequences of sight deprivation are vast and usually ignored. Blindness has an impact on emotional status, self-identity, and social interaction. The visually impaired often feel isolated, frustrated, and helpless, especially when there is a sudden or progressive loss of vision (Boagy et al., 2022). The most common among the people with visual disabilities are depression and anxiety because the lack of independence, fear of further decline, and social exclusion cause depression and anxiety among people with visual impairment to a greater extent than in the general population. Social participation disability is experienced as a result of face recognition, non-verbal interpretation, and interaction with a group. Such problems may result in social withdrawal, less interaction in the community, and poor interpersonal relations. Stigma and negative attitudes of society toward disability also enhance exclusion, supporting internal feelings of inadequacy and marginalization (Kumar et al., 2024). In children, vision impairment may lead to developmental disabilities in terms of motor, cognitive, and social development. Poor visual stimulation at the critical stages of development slows down motor coordination, spatial awareness, and social learning. These delays may continue until adulthood, leading to compounded disability in more than one area unless early intervention is applied. Vision-related disability may be cumulative and lifelong in cases where there is no intervention and provision of supportive conditions. Functional limitations, psychological distress, social exclusion, and economic hardship have synergistic interactions that lead to multidimensional disability profiles (Akyol Güner & Das Gecim, 2023). The cycle needs to be stopped so that individuals with vision impairment can enjoy the best participation and quality of life by undergoing an extensive rehabilitation process, inclusive policies, available technologies, and the acceptance of these individuals in society.

5.3 Deafness and Hearing Disability

Hearing is a basic sensory activity, and it is at the core of human communication, language acquisition, socialization, education, and awareness of their surroundings. Deafness and hearing impairment are serious sensory disorders that are capable of causing profound and lifelong disabilities if not mitigated properly (Shodieva, 2025). Hearing loss is an invisible disability compared to most physical impairments, which are immediately evident; therefore, though its functional effects are not visible, it is equally greatly affected. Hearing disability is a disability that spans across all age groups, including congenital hearing loss in infants and age-related hearing impairment in the elderly, but the impact of hearing disability goes much further than the hearing system to affect cognitive, emotional, educational, and socioeconomic levels. Deafness is a very severe or profound loss of hearing whereby auditory perception is insignificant or missing even after amplification (Rani et al., 2025). The hearing impairment, however, is a bigger range of hearing sensitivity damage, from mild to severe loss. Whereas medical definitions lay emphasis on the auditory thresholds, disability arises when hearing impairment disrupts the smooth functioning of communication, involvement in daily activities, and social and occupational integration. Modern disability theories suggest that hearing disability is based on the interplay of impaired hearing and situational factors, including the absence of effective communication mechanisms, inadequate rehabilitation, stigma, and insufficient policy provision (Aldè et al., 2025). The global population has been aging, there has been exposure to noise in the environment, untreated ear diseases, and poor access to early screening and intervention in most regions, which makes the burden of hearing disability to be on the increase. Most notably, hearing disability is not only a health problem but also a developmental, educational, and human rights issue, and it should be addressed on a multidisciplinary and integrated basis.

5.3.1 Causes of Hearing Impairment

Hearing loss can either be inborn or acquired, and it can happen at any age. Congenital deafness is commonly linked to genetic prenatal infections, birth complications, or ototoxic exposure in pregnancy. In children with early hearing loss (before the learning of language), it may profoundly impair speech and language development, leading to communicative disability in the long term (Lamminmäki et al., 2023). When this is so, the disability is not restricted to auditory perception but spreads to the linguistic, cognitive, and social levels. The causes of acquired hearing impairment include frequent ear infections, trauma, excessive noise exposure, ototoxic drugs, or degenerative changes with age. One of the most widespread sensory disorders in the world is age-related hearing loss, also known as presbycusis, which tends to develop over time so that a person may not consider its effects to be harmful until serious disability sets in (Yang et al., 2023). The progressive hearing loss is quite debilitating since a person might find himself or herself becoming progressively unable to comprehend speech, particularly in areas with a lot of noise, way before he or she becomes totally deaf. The effects of impaired hearing that are functional in nature are determined by a number of factors, such as the severity, early or late onset, the duration of the hearing impairment, the access to assistive equipment, and the communication environment around the affected individual. Hearing impairment easily becomes a disability in societies where oral communication is prevalent and captioning or similar service provision is minimal (Zárate, 2021). Therefore, the hearing disability is not only conditioned by the biological impairment but also by the social context and the technological context.

5.3.2 Communication and Language Disability

The most characteristic and debilitating outcome of deafness and hearing loss is communication disability. The impaired hearing directly disrupts the process of hearing spoken language, which is the major form of communication in most societies. Hearing-impaired people tend not to listen to speech in a clear way, especially in group discussions, crowded places, classrooms, or workplaces where there is noise (Akay, 2023). Misunderstandings, slow responses, and communication failures are common, leading to frustration for both the individual and their communication partners. In the case of those deaf since childhood or early childhood, the lack of sound input may be very detrimental to the normal course of speech and language development unless the condition is treated at an early age. Early childhood language acquisition largely relies on hearing exposure, and any delays in hearing could lead to a lifelong language impairment. These language impairments can disrupt reading, writing, abstract thinking, and academic performance, potentially leading to long-term educational disabilities. Communication disability may also be experienced in people with partial hearing loss, as there will be an added effort in listening, fatigue, and a lower level of understanding (Davis et al., 2021). Lip-reading, context, and repetition, which many persons with hearing impairment use, are cognitively challenging and unreliable. In the long run, this constant struggle of communication may cause mental strain, disengagement with the dialogue, and shying away from social interaction. The environmental barriers to communication disability include the unavailability of sign language interpreters, non-capturing media, and ignorance by hearing people. People with hearing impairment can easily be locked out of education, health care discourse, law making, and civic life in the absence of accessible communication accommodations (Harris & Tani, 2021).

5.3.3 Educational and Social Participation Disability

The impact of deafness is very widespread in education, employment, and social life. The obstacle faced by children with hearing impairment in the learning institutions is almost always enormous due to inadequate infrastructures of an inclusive nature, unqualified teachers, and a lack of assistive technology. In a classroom environment, learning is usually sound-based, and unless hearing-impaired students are accommodated, such as through interpreting sign language or using captioning or assistive listening devices, learners may not be able to follow lessons, participate in a discussion, and achieve academic achievements (Adeduyigbe et al., 2024). Such problems can result in poor levels of education, poor literacy levels, and even school dropouts. Auditory disability is a severe problem which affects the workforce and the employment of adults. Many of these jobs require that one use verbal communication appropriately, work in a team, and have a sense of hearing. Misconceptions about the abilities of individuals with hearing impairment may subject them to discrimination, a lack of employment opportunities, and a lack of career progression because of false beliefs about their abilities. Inaccessible work environments and unreasonable accommodations may limit productivity and participation even in employed individuals and result in underemployment or job dissatisfaction (McArthur et al., 2025). The other significant impact of hearing impairment is social participation disability. The inability to keep up with conversations, particularly in social events, may result in embarrassment, isolation, and withdrawal. Patients can skip social responses, community, and socialization to stem social communication pressure. Eventually, this shrinkage of social networks and support systems makes people more exposed to loneliness and marginalization. The compounding influence of educational, work-related, and social exclusion supports the disabling influence of hearing impairment. Hearing-disabled people are methodically discriminated against in various aspects of life without comprehensive policies and available environments (Tripathi & Saranya, 2022).

5.3.4 Cognitive and Long-Term Disability

The psychological effect of deafness and hearing impairment is quite significant and is underrated most of the time. Depression, anxiety, and emotional distress are more likely to affect people with hearing disabilities than they are to affect the rest of the population (Cejas et al., 2021). There is the existence of communication barriers, social isolation, and perceived stigma, thus resulting in low self-esteem and inadequacy. To those who become deaf later in life, adjusting to this new fact may be especially difficult. These individuals are forced to confront the loss of a sense they relied on and to learn new ways of communicating. It has also been attributed to cognitive impairment, particularly among the elderly due to hearing impairment (Powell et al., 2021). Before treatment, it has been found that hearing loss has caused a quick deterioration in cognition and dementia. Some of the processes proposed by researchers are low stimulus levels in the auditory system, high levels of cognition in listening, and social isolation. Thus, hearing disability is not just a sensory and communicative dysfunction issue but also a problem that has a broader scope of consequences on the health and cognition of the brain. Hearing impairment in children can lead to prolonged hearing loss, which in turn results in impaired cognitive functions, executive skills, and social cognition, particularly in cases where there has been a delay in language exposure (Lauriello et al., 2024). These effects can persist into adulthood, impacting education, employment, and social self-sufficiency levels. Typically, the severity of hearing disability increases over time. Social withdrawal that leads to psychological distress and mental degeneration can also be caused by difficulties in communication (Koszalinski & Olmos, 2022). There is also economic dependency and restricted access to services, which aggravates the disability. Without early diagnosis and treatment, as well as community support, hearing loss can develop into a complex and lifelong disability that affects nearly all aspects of human functionality.

5.4 Tactile Impairment

The human being possesses a sense of touch that is essential for interacting with the physical world. By touch, a person is able to feel the texture, temperature, pressure, vibration, pain, and proprioceptive signals as to the position and movements of the body. Such sensations are a component that is vital in the execution of daily activities, safety, control of motor abilities, and emotional and social interaction. Tactile sensation impairment is thus a critical sensory impairment disorder with far-ranging implications that are not limited to sensory loss alone but tend to lead to a lot of disability and limitation of participation (Zhang et al., 2022). Tactile impairment is a partial or total loss, distortion, or abnormal sensation of touch-related sensations. The impairment can be caused by peripheral nerve damage, central nervous system disorders, metabolic diseases, trauma, infections, or degenerative conditions. In comparison to visual or auditory impairment, tactile impairment is less apparent and thus poorly recognized, and it is not timely treated (Xu et al., 2024b). However, it may have an extensive effect on functional autonomy, working capacity, mental health, and quality of life. Tactile impairment disability occurs in cases where sensory impairment disrupts the normal capacity to execute basic functions or engage in social, educational, and economic activities in their entirety. Tactile impairment can be viewed as an impairment at the level of body function in relation to the International Classification of Functioning, Disability, and Health (ICF), and, as such, is the interaction between environmental and personal factors and activity limitations and participation restrictions (Jaiswal et al., 2024). Appreciating tactile impairment in this type of disability approach is essential to a holistic approach to rehabilitation and inclusive policy making.

5.4.1 Physiology of Tactile Impairment

The tactile sensory system is run by special mechanoreceptors that are found in the skin, muscles, joints, and connective tissues. These receptors also pass information through the peripheral nerves to the spinal cord to the brain, where sensory information is combined and decoded (D'Mello & Dickenson, 2008). Proper tactile feedback is required to facilitate fine motor skills and control, manipulation of objects, protective reflexes, and body localization. When any of the parts of this sensory route are impaired, it results in tactile impairment. Peripheral neuropathy, spinal cord injury, stroke, multiple sclerosis, diabetic neuropathy, chemotherapy-induced neuropathy, leprosy, and nerve damage from trauma are common causes. The central processing disorders can lead to a change in tactile discrimination despite intact peripheral sensation. Tactile impairment is, in most instances, accompanied by pain syndromes like neuropathic pain and makes functional outcomes more difficult (Auld et al., 2014). The loss of the tactile feedback interferes with the manner in which the brain regulates the motor output. Consequently, people can find it difficult to perform precise tasks, balance, grip strength, and postural control. These physiological changes form the basis for the development of disability.

5.4.2 Functional Disability Due to Tactile Impairment

One of the most direct effects of loss of the sense of touch is functional disability. Touch is a major activity in activities of daily living that involves handling of objects, force modulation, and being aware of the environment (Sonneveld & Schifferstein, 2008). Where there is a loss or lack of tactile feedback, the individual experiences constant challenges in most of their daily activities. Task performance like buttoning clothes, tying shoes, writing, cooking, or operating tools, becomes difficult as people are unable to estimate the pressure or position of objects properly. This usually causes a lot of dropping of objects, inefficient movement, and time spent doing the work. In extreme situations, visual compensation can cause dependence in individuals and cognitive load and fatigue. Tactile impairment also disrupts self-care activities like bathing and grooming (Waugh, 2013). Decreased sensation can mean that people cannot perceive the temperature of the water, which exposes them to burns or scalds. Equally, lack of pain or injury sensation can lead to the fact that there can be unnoticed wounds, especially when diabetes has been diagnosed with a neuropathy and causes infections, ulcers, and, in the worst scenario, amputations (Hillson, 2017). These secondary complications are a big burden on disability and healthcare.

5.4.3 Mobility and Postural Disability

Balances and coordinated movements depend on tactile feedback from the feet, joints, and muscles. Touch and proprioception impairment interferes with gait stability and postural control. Tactile-impaired persons tend to have unsteady patterns of walking, a lack of coordination, and high dependency on visual feedback to keep balance (Rana et al., 2025). With the reduction of tactile feedback, the presence of falls can be viewed as a significant threat, especially in elderly patients and those with neurological problems. Falls tend to cause fractures, head injuries, and long-term mobility restrictions and turn a sensory impairment into a permanent physical disability. The fear of falling is an additional limitation of movement that causes a lack of physical activity, muscle weakness, and social isolation. Tactile impairment can cause mobility disability, thus severely limiting independence (van Dam et al., 2024). It becomes dangerous to navigate uneven surfaces, climb up stairs, or walk under low light conditions. The people can, over time, limit their mobility to the known areas, further promoting isolation and reliance.

5.4.4 Psychological and Emotional Disability

The psychological impacts of the loss of the sense of touch are not often considered. A loss of senses alters the perception of the body by people and the way they interact with the environment, which can have significant effects on emotional health and even self-identity. The resulting tactile impairment, which develops chronically, is associated with an increase in anxiety and depression, particularly in situations where it leads to loss of independence or even employment (Roby-Brami et al., 2021; Ståhl et al., 2023). Frustration, helplessness, and fear of getting hurt or falling ill can be developed by people. Mental exhaustion and emotional suffering are supplemented by the constant requirement to remain alert to compensate for the loss of senses. Touch is also significant in emotional regulation and social attachment. The disrupted sense of touch may reduce the emotional enjoyment of physical contact, such as holding hands or hugging each other, which results in the feeling of being isolated and being alone. The existence of tactile impairment may also interfere with emotional development and attachment in children, and this may be the only contributor to psychological disability (Sahoo et al., 2021; Fernandez et al., 2025). Social participation may be limited to a great extent by tactile impairment. Difficulties in executing social gestures or attending to group or recreational events reduce the chances of interaction (Tseng et al., 2022). Isolation may occur as people fear being hurt or humiliated and therefore shun social places. Another element that causes barriers to participation is environmental inaccessibility. Even community facilities, transportation, and communal places are usually not built taking into consideration the requirements of the visually and hearing-impaired individuals, and thus, including them is even more difficult. There is also a societal lack of awareness regarding tactile disabilities, and this leads to stigma and misinterpretation, hence enhancing exclusion. Social disability has an even greater impact when it is combined with tactile impairment and other sensory or physical impairments (Philip & Santhosh, 2025). A combination of several impairments can have a cumulative effect that results in severe restrictions to participation and reduced quality of life.

5.4.5 *Developmental Disability in Children with Tactile Impairment*

Tactile sense is vital in the normal motor, cognitive, and social development of children. Premature tactile impairment may cause dysfunction in the sensory integration processes, resulting in delayed motor milestones, poor coordination, and loss of spatial awareness. Children with tactile impairment can show difficulties in handwriting, manipulating objects, and playing (Park et al., 2021). These operational constraints may have impacts on schoolwork and interpersonal association. In case of the lack of early intervention, there is a chance of the emergence of secondary developmental disabilities such as decreased self-confidence, behavioural problems, and learning disabilities. Early exposure and occupational therapy help to reduce long-term disability among children (McQuiddy et al., 2024). Tactile impairment is a problem that may lead to adverse functionality and participation throughout a person's life if it is not detected and managed at early stages of development.

5.5 Gustatory Impairment

The gustatory system is a vital part of human survival, food, pleasure, and social interaction. Taste perception helps individuals in recognizing nutrients, preventing toxins, controlling appetite, and finding pleasure in food and drinks (Hossain et al., 2025). It has a complex connection with olfaction, somatosensory information, and higher cortical processing, so it is a multisensory experience, not an isolated activity. In case of gustatory perception impairment, the impairment does not just end with a loss of taste sensation; it is, in most cases, followed by a great functional disability, nutritional imbalance, psychological distress, and impaired quality of life. Gustatory impairment is the impaired or distorted perception of taste to a certain degree. It has such conditions as hypogeusia (loss of taste sensitivity), ageusia (absolute loss of taste), dysgeusia (alteration of taste perception or unpleasant taste perception), and phantogeusia (taste perception without stimulus). Taste impairment is a significant sensory disability, though it is commonly not given serious consideration in comparison to vision or hearing loss, especially when chronic or severe (Okada et al., 2021; Molinari et al., 2022). Its effects are seen to be more severe when it comes along with other sensory impairments, systemic health conditions, old age, or neurological conditions. Disability-wise, gustatory impairment is not just a sensory anomaly but a disorder disrupting normal daily functioning, engagement in social and cultural activities, nutritional health, and emotional well-being. According to the International Classification of Functioning, Disability and Health (ICF), sensory impairments are seen as a source of disability where they prevent activities or limit activities (Cozzi et al., 2021). This category fits taste impairment well, especially when environmental and individual factors intensify the effects.

5.5.1 Pathophysiology of Gustatory Impairment

The gustatory system is highly complex since both peripheral and central factors can cause an impairment in the sense of taste in a wide range of possible ways. Examples of peripheral causes include the destruction of taste buds, salivary gland dysfunction, oral problems, dental problems, and medication-induced changes. Lesions or dysfunction of neural pathways processing taste, including cranial nerves VII (facial), IX (glossopharyngeal), and X (vagus), and cortical and subcortical brain areas, are the causes of the problem that may be central (Udagatti et al., 2023). Gustatory dysfunction is a usual feature of stroke and Parkinson's, Alzheimer's, multiple sclerosis, and traumatic brain injury (Kushwaha et al., 2023). In such instances, the impairment of taste can be observed together with the impairment of other senses or motor deficits, which increases overall disability. Systemic diseases like diabetes mellitus, chronic kidney disease, liver disease, and endocrine abnormalities also change the perception of taste using metabolic and neural pathways (Kushwaha et al., 2023). Another important cause is medical intervention. Taste disturbances are typical of chemotherapy, radiotherapy (particularly in the case of head and neck cancers), and the use of long-term medications. Treatment-induced impairments often go unnoticed, despite their substantial impact on nutrition and treatment adherence. Older adults often experience age-related degeneration of taste buds and neural pathways. This decline typically occurs alongside a diminished sense of smell, dry mouth (xerostomia), and the use of multiple medications. Gustatory impairment can continue to exist despite the corrective treatment of the underlying cause and can thus turn a temporal sensory disability into a permanent disability (Putri & Astiarani, 2023). The absence of efficiency in the restorative therapies implies that a significant number of people have to cope with the long-term loss of taste, and the success rates of it are different and may rely on individual resistance and environmental assistance.

5.5.2 Functional Disability and Nutritional Consequences

Gustatory impairment is one of the most profound disabilities in its influence on nutrition and eating behaviour. Taste is a crucial aspect of appetite control, food choice, and satisfaction with a meal. In case of lost or altered taste, the person could lose interest in food, and food consumption will decrease, causing unbalanced food intake or avoidance of food altogether. It may cause weight loss, malnutrition, deficiencies of micronutrients, and a compromised immune system, especially in the elderly and those with chronic diseases (Younes, 2024). On the other hand, others use low taste to balance the lack of taste by adding more salt, sugar, or spices to make food taste good, which could exacerbate diseases like high blood pressure, diabetes, and heart disease. This secondary compensation explains the way in which gustatory impairment may be an indirect cause of secondary health disabilities (Oleszkiewicz et al., 2023). Taste impairment can interfere with the normal feeding patterns and development in children, resulting in developmental and nutritional difficulties. The other impairment, related to food safety, is taste. Failure to identify spoiled or contaminated food is one of the factors that pose a risk of foodborne illness. People with a complete loss of taste will not be aware of the fact that they might be taking dangerous substances, especially when their sense of smell is also lost. The protective role of taste, along with the safety-related disability related to its impairment, is emphasized by this functional restriction (Zhou et al., 2024). In addition to nutrition, gustatory impairment influences adherence to medications. Many oral drugs are not palatable, and people with dysgeusia cannot tolerate medications, thus missing doses or terminating the use of the drug. This has a trickle-down effect of health complications, which worsen disability burden.

5.5.3 *Psychosocial Dimensions of Taste-Related Disability*

The experience of taste has strong connections with emotional experience, memory, and social interaction. Cultural identity, bonding in the family, celebrations, and rituals all revolve around food. Gustatory impairment intrudes upon such experiences, with frequent psychosocial effects that are devastating. People often complain of experiencing some sense of loss, grief, or withdrawal of the positive elements of life after they lose their sense of taste. Depression, anxiety, and lack of satisfaction with life are closely related to chronic taste disorders. Failure to enjoy food reduces daily pleasure, and withdrawal during socialization (sharing food) may be a result of embarrassment or not feeling like being together with the food (Pineau et al., 2021). This may eventually result in isolation, less social engagement, and lower emotional status, especially among elderly people who live singly. Distortions in the sense of taste, including the continued sense of metallic taste or bitter taste, are particularly uncomfortable. Dysgeusia can lead to nausea, food aversion, and persistent discomfort, which has a serious adverse effect on mental wellness (Jafari et al., 2021). There are occasions when people become conditioned to avoid foods that they once liked, and this further reduces the food diversity and social interaction. The sense of identity and independence can also be impaired as a result of taste. Persons formerly fond of cooking, eating, or work involving food (as chefs or employees in the food industry) can feel jobless and disabled. This is an example of the way gustatory impairment is not visible but can potentially lead to life-changing effects, just as other sensory impairments of a more familiar nature.

5.5.4 *Participation Restriction*

In terms of participation, gustatory impairment reduces the ability to engage in social, cultural, and occupational activities revolving around food (Fjaeldstad & Smith, 2022). Communal meals, religious practice, celebrations, hospitality, and so forth usually lose their meaning or turn out to be frustrating. The participation limits are compatible with the ICF model, according to which the manifestation of disability is in the interaction of sensory disability and social expectations. Irrespective of its effects, gustatory impairment is a poorly understood phenomenon in disability policy, clinical treatments, and rehabilitation. Taste disorders do not often need institutional benefits and programmed rehabilitation, unlike vision or hearing impairment. This omission increases the unmet needs and incomplete support of the affected individuals. Gustatory impairment rehabilitation is mainly concerned with adaptation and not restoration (Yu et al., 2024). Disability can be alleviated by means of nutritional counselling and sensory retraining, sensation-enhancement techniques, and psychological support, which can also enhance the quality of life. The secondary risks need to be reduced by educating patients on safe food practices and medication management. Multidisciplinary interventions such as the integration of physicians, dietitians, psychologists, and speech-language or sensory therapists have the highest results, especially in individuals with intricate or prolonged impairments (Haanes, 2023). Inclusive healthcare and disability systems should be more aware of gustatory impairment as a valid sensory disability. With growing populations living with age, the drawbacks associated with age, like a higher number of people suffering from chronic illness and medical procedures, will move to taste-related disability. Their identification of functional, psychosocial, and participatory outcomes is a vital measure in all-encompassing sensory disability care.

5.6 Olfactory Impairment and Associated Disability

The olfaction, or sense of smell, is a vital but undervalued sense that contributes to the survival, safety, nutritional, emotional, and social aspects of human life in a primary way. Olfactory perception helps a person to sense any hazards in the environment, like smoke vapours, gaseous leaks, bad food, and poisonous chemicals, as well as playing an important role in flavour perception, appetite control, memory creation, and emotional states. Olfactory impairment is a partial or total loss or distortion of the sense of smell, which results in such conditions as hyposmia (reduced smell), anosmia (total loss), parosmia (distorted smell perception), and phantosmia (perception of odours with no stimuli) (Schäfer et al., 2021). Although the role of olfactory disorders might be less debilitating in comparison to vision or hearing loss, their effect on everyday activities, mental health, and quality of life is significant and usually underestimated. The impairment of the olfactory ability can be a result of abnormalities during development, infection of the upper respiratory system, neurodegenerative diseases, brain damage, ageing, contact with toxins, or chronic sinus and nasal diseases (Fatuzzo et al., 2023). In recent years, the phenomenon of the disabling impact of the loss of smell has attracted the attention of various countries to post-viral olfactory dysfunction, especially after becoming infected with COVID-19. Olfactory disorders are often non-evident and poorly identified, which is unlike other sensory disabilities and leads to poor diagnosis and rehabilitation as well as insufficient social support (Hummel et al., 2023). The loss of smell can become a serious sensory impairment when it affects the ability to engage in productive activities, personal safety, nutrition, emotional stability, and social participation.

5.6.1 Functional Disabilities from Olfactory Impairment

Environmental awareness and perception of safety is one of the most immediate and severe effects of olfactory impairment. People with a sense of smell perceive dangerous smells like smoke, gas leaks, burning wires, or chemical fumes. This greatly exposes people to domestic accidents, poisoning, and life-threatening conditions. Olfactory dysfunction may limit work or career modification within the workplace, especially for those who involve chemical handling, food processing, or environmental monitoring and surveillance, and hence functional disability. Another dangerous problem that affects both nutrition and diet is olfactory dysfunction (Melis et al., 2023). Smell is critical in flavour, and its loss results in a bad taste sensation, despite the normal gustatory functioning. Food may be bland, unpleasant, or unrecognizable, and it may lead to the reduction of appetite, altered food behaviour, or excess salt, sugar, and spices to balance the loss of sense of taste. Such impairments can ultimately lead to malnutrition, undesired loss or gain of weight, and metabolic imbalance. The risk factors that are connected with olfactory dysfunction in older adults are noted as frailty, nutritional deficiency, and high mortality rates, therefore, demonstrating the influence of olfactory dysfunction on functionality and health-related disability (Yeo et al., 2024). Domestic management and personal hygiene are also affected by the activities of daily living. Olfactorily dysfunctional patients cannot sense body smells, spoiled food, contaminated environments, and smoke, which causes them to have poor self-care and hygiene. This can result in embarrassment, social withdrawal, or relying on other people to assure them that they are clean and safe. Such practical constraints depict the way in which the loss of olfaction not only goes beyond the sensorineural loss but also encroaches on autonomy and dignity.

5.6.2 Psychosocial Disability Associated with Smell Loss

The psychosocial implications of olfactory impairment are significant and very frequently ignored. Smell is closely associated with the limbic system, which controls memory and emotions. Olfactory loss interferes with the process of emotion and suppresses the intensity of positive experiences attached to food, nature, relationships, and personal identity (Blomkvist & Hofer, 2021). People commonly complain of emptiness, detachment, and emotional numbness after the loss of smell, which may advance to depressive symptoms. It also influences social interaction. Smell has a lesser and a significant role in social bonding, interpersonal attraction, and emotional communication. Olfactory stimuli determine maternal attachment, love, and comfort in interpersonal relations. Olfactory-impaired individuals can experience feelings of being out of touch with others or fearful of being rejected by society because they are unable to detect either personal or environmental scents (Marin et al., 2023). The fear of accidentally releasing offensive smells may lead to increased shyness and antisocial behaviour, which supports isolation. Psychological disability becomes more severe with the sudden or unexplained loss of the sense of smell. A sudden loss of sensory experience may lead to anxiety, grief, and dissociation of identity. Olfactory loss has been defined by many people as the loss of a part of the self when memory of smell and emotional feelings become inaccessible (Diaconu, 2021). Conditions such as parosmia and phantosmia may arise, which involve experiencing false scents. Such conditions may result in symptoms such as nausea, aversion to food, and constant psychological suffering. If they don't receive the necessary counselling and support, these emotional implications could eventually lead to mental health disabilities.

5.6.3 Participation in Social Disability

Smell disability becomes one of the key factors contributing to the restriction of involvement in social, cultural, and working settings. Food, common meals, and indulgence in sensual pleasure are the focus of social activities (Błochowiak, 2022). Individuals with no sense of smell are then out of this type of experience since they cannot discuss taste, smell, and the enjoyment of food. This lack of connection may lead to a reduction in family activities, partying, and cultural activities, particularly in those cultures where food and smell are symbolic. In the employment environment, a lack of smell may limit the number of roles an individual can perform in the food and beverage industry, perfumery, food quality inspection, medicine, firefighting, and environmental safety. The fear of safety, job performance, or stigma may affect the career advancement opportunities, even when these careers do not necessarily depend on smell. Lack of workplace accommodations and low levels of employer awareness exacerbate employment-related disabilities. Social disability also contributes to the challenges faced by individuals with olfactory impairment due to its invisible nature (Fatuzzo et al., 2023). Unlike visual and hearing impairment, smell loss has no external signs, so others may misinterpret and downplay it. People's lack of acceptance makes it challenging to justify their deprivation or obtain disability assistance. This invisibility can contribute to marginalization and reduced access to social resources, which promote constraints to participation and inequity.

5.6.4 *Olfactory Impairment in Ageing, Disease, and Long-Term Disability*

Closely related to the ageing process and neurodegenerative disorders such as Parkinson's and Alzheimer's, as well as other forms of dementia, is the close correlation between ageing and olfactory impairment. The loss of smell is, in most instances, a prodrome of an imminent neurological deterioration that usually comes along with cognitive and motor phenomena (Willem, 2022). When these diseases develop, they interrelate to impairment of smell, and therefore, due to cognitive dysfunction, they become more reliant and require more care. The loss of the ability to smell, see, and hear in older adults combined together to form combined sensory impairment, leading to the compounded disability, which significantly affects the quality of life and survival. Long-term health monitoring and disease control are also other implications of the chronic loss of olfaction. The deficit in the ability to smell body smells may delay the mechanism of detecting infection, injuries, or metabolism (Matiashova et al., 2023). The lack of appetite and enjoyment in the eating process can worsen the outcome of chronic disease. In addition, chronic olfactory impairment has also been linked to increased vulnerability to depressive emotions, weakness, and demise, a factor that demonstrates its importance in its ability to be a predictor of general health vulnerability. Despite the effect, there are limited rehabilitation choices for olfactory impairment. Important parts of the management that have been used but infrequently in the healthcare systems are safety education, nutritional guidance, counselling, and olfactory training. There are also deficits in standardized assessment measures and classifying disability, which are also contributing to the factors of underdiagnosis and poor support. It is important to note that olfactory impairment is a valid sensory disability that needs to be considered in order to establish inclusive policies and facilitate research and the provision of rehabilitation services (De Luca et al., 2023).

5.7 Challenges Faced by Individuals with Sensory Impairment

The sensory impairments, such as vision, hearing, tactile, olfactory, and gustatory disorders, present complex problems that go much beyond the absence of sensory capabilities. The issues can impact almost all areas of human life, including the ability to survive and stay safe, the ability to socialize, mental health, school, and financial self-sufficiency. The extent to which a person with sensory impairment is disabled is not dependent on the severity of the sensory loss alone, but rather on a combination of the biological restrictions, the physical and social environments, the attitudes of society, and the support systems (Armstrong et al., 2022). Since perception, learning, communication, and interaction with the external world are based on the perception of the senses, their deficiency interferes not only with personal autonomy but also with social integration.

5.7.1 Daily Living Challenges

Problems in carrying out activities of daily living are one of the most imminent and chronic problems experienced by someone who has sensory impairment. Eating, dressing, personal hygiene, navigation, and household control are only some of the functions that require sensory information. The impairment of vision may support the inability to identify objects, define the spatial orientation, and move safely; the impairment of hearing can influence the possibility to identify the alarms, verbal instructions, or environmental sounds. Tactile impairment lowers sensitivity to temperature, pressure, and pain, making one susceptible to burns, injuries, and hidden wounds. Smell and taste disorders disrupt food identification, appetite control, and the smell of dangerous substances like smoke or escaping gas (Carvill, 2001; Rajput et al., 2023). With the accumulation of these sensory impairments, people tend to become less independent and more dependent on helpers or assistive technologies. This dependence can be partial or complete based on the levels of impairment of the senses and their combination. Eventually, this dependency on others in accomplishing daily tasks may cause frustration, a lack of confidence, and a breakdown of self-esteem. In elderly patients, impairment of sensory functions is often a rapid accelerant of functional impairment, which in turn leads to frailty and institutionalization. In children, deficient senses could ensure slower mastering of self-care skills, which restricts autonomy in adulthood.

5.7.2 Social Interaction Challenges

Sensory perception, especially hearing and sight, is an essential part of effective communication. The hearing impairment poses a high obstacle to verbal communication, which in most cases leads to misunderstanding, lack of involvement in conversations, and withdrawal. Deaf people might find it hard to fit in a social, educational, and professional sphere that does not provide them with sign language interpretation or captioning, which results in their rejection by others, (Zárate, 2021) In the same way, vision impairment also influences nonverbal communication because it reduces the capacity to interpret facial expressions and gestures as well as other forms of social interaction, which are vital aspects of interpersonal relationships. The lack of sensory sensation will make social interactions more complicated because, contrary to the situation when a person has both senses, the person will face embarrassment, perceived incompetence, or frustration during the interaction (Sterkenburg et al., 2022; Tam & Koppel, 2021). Negative experiences conducted on a regular basis may eventually demotivate individuals to start or continue social interactions. This isolation is also heightened by the attitudes of the society that stigmatize disability, which tend to perceive the sensory impairment in terms of pity or incapability and not disability. Consequently, people can absorb negative attitudes, which strengthen withdrawal and restrict social involvement.

5.7.3 Educational Challenges

The impairments of the senses are a great barrier to both learning and work contexts, where the major mode of delivering information is by sight and sound. Visual or hearing learners frequently encounter unaccommodating learning resources, untrained teachers, and inadequate facilities. Lack of Braille materials, assistive hearing devices, captioned media, or tactile learning materials hinders learning excellence and academic achievement (Zárate, 2021; Tripathi & Saranya, 2022). The sensory barriers may cause underachievement or a lack of the ability to learn or drop out, even in cases where intellectual ability is not compromised. These educational gaps in adulthood are converted into low employment and economic insecurity. Most workplaces are not equipped well to accommodate sensory-impaired employees and are not fitted with accessible communication networks, adaptive technology, or inclusive policies. People can be locked out of some careers because of prejudice over their capabilities instead of the real functional impairments. This has led to a higher rate of unemployment and underemployment among the sensory disabled (Ghazy et al., 2025). Economic dependency also contributes to disability by limiting access to healthcare, rehabilitation services, and assistive technologies due to the cycle of disadvantage that is hard to overcome.

5.7.4 Emotional Challenges

The psychological effect of sensory deprivation is immense and is usually not taken seriously. The loss of senses can significantly impact an individual's sense of identity, particularly when the impairment occurs later in life. There is a possibility of sudden or progressive loss of vision, hearing, or touch as the grieving process, which includes denial, anger, sadness, and adjusting to the situation. Situations of anxiety are typical, particularly when there is uncertainty in the areas of safety, mobility, or communication (Sharpe et al., 2022; World Health Organization, 2022). Sensory impairment often goes hand in hand with depression, and it is motivated by isolation, loss of independence, and reduced quality of life. Sensory-impaired children can find it difficult emotionally because of slow social development, bullying, or exclusion. Teenagers usually have a problem with self-esteem and acceptance by their peers, whereas adults can lose self-confidence in their work and personal lives. Older adults experience compounded psychological stress due to the combination of sensory impairment, aging-related decline, chronic illness, and bereavement. Such emotional problems may develop into psychological disability in the long term without proper mental health assistance, and they continue to restrict participation and well-being (Gorur & Dey, 2021; Cote, 2021).

5.8 Conclusion

The problems of the people who have sensory impairment are multidimensional, interdependent, and lifelong. The lived experience of sensory disability is influenced by functional limitations, communication barriers, educational and occupational barriers, psychological distress, and environmental barriers. Social structures, societal attitudes, and policy decisions largely shape these problems, not the inevitable results of sensory impairment. The sensory impairment has to be addressed through an all-inclusive strategy that is not only based on clinical treatment but also upon the realization of design, technology, inclusive education, employment equity, and mental attention. This approach can identify and fully solve these problems, leading to disability mitigation, improved involvement, and the dignity and rights of persons with sensory impairments in society.

References

Adeduyigbe, A. M., Adeduyigbe, A. E., & Tijani, B. E. (2024, December 1). Addressing students with hearing impairment's current state and future needs: Reforming an inclusive science education. *International Journal of Studies in Inclusive Education, 1*(2), 12–15.

Akay, E. (2023, January). Teachers' opinions on the attending of the hearing-impaired students in the inclusion environment and the resource room services. *Journal of Qualitative Research in Education, 4*(33).

Akyol Güner, T., & Das Gecim, G. Y. (2023, August). Effects of social exclusion on psychological-well-being and suicidal possibilities among people with physical disabilities. *OMEGA-Journal of Death and Dying, 87*(3), 962–976.

Aldè, M., Ambrosetti, U., Barozzi, S., & Aldè, S. (2025, April 24). The ongoing challenges of hearing loss: Stigma, socio-cultural differences, and accessibility barriers. *Audiology Research, 15*(3), 46.

Alotaibi, H. M., Alduais, A., Qasem, F., & Alasmari, M. (2025, February 24). Sensory disorders and neuropsychological functioning in Saudi Arabia: A correlational and regression analysis study using the National Disability Survey. *Healthcare, 13*(5), 490.

Armstrong, N. M., Vieira Ligo Teixeira, C., Gendron, C., Brenowitz, W. D., Lin, F. R., Swenor, B., Deal, J. A., Simonsick, E. M., & Jones, R. N. (2022, July). Associations of dual sensory impairment with incident mobility and ADL difficulty. *Journal of the American Geriatrics Society, 70*(7), 1997–2007.

Auld, M. L., Russo, R., Moseley, G. L., & Johnston, L. M. (2014, September). Determination of interventions for upper extremity tactile impairment in children with cerebral palsy: A systematic review. *Developmental Medicine & Child Neurology, 56*(9), 815–832.

Benavot, A., Hoppers, C. O., Lockhart, A. S., & Hinzen, H. (2022, April). Reimagining adult education and lifelong learning for all: Historical and critical perspectives. *International Review of Education, 68*(2), 165–194.

Błochowiak, K. (2022, September 30). Smell and taste function and their disturbances in Sjögren's syndrome. *International Journal of Environmental Research and Public Health, 19*(19), 12472.

Blomkvist, A., & Hofer, M. (2021, January 1). Olfactory impairment and close social relationships. A narrative review. *Chemical Senses, 46*, bjab037.

Boagy, H., Jolly, J., & Ferrey, A. (2022, November 10). Psychological impact of vision loss. *Journal of Mental Health and Clinical Psychology, 6*(3).

Boldsen, S. (2022, July 25). Autism and the sensory disruption of social experience. *Frontiers in Psychology, 13*, 874268.

Carvill, S. (2001 Dec). Sensory impairments, intellectual disability and psychiatry. *Journal of Intellectual Disability Research, 45*(6), 467–483.

Cejas, I., Coto, J., Sanchez, C., Holcomb, M., & Lorenzo, N. E. (2021, April 1). Prevalence of depression and anxiety in adolescents with hearing loss. *Otology & Neurotology, 42*(4), e470–e475.

Cote, A. (2021, December 1). Social protection and access to assistive technology in low-and middle-income countries. *Assistive Technology, 33*(sup1), S102–S108.

Cozzi, S., Martinuzzi, A., & Della, M. V. (2021, December 29). Ontological modeling of the international classification of functioning, disabilities and health (ICF): Activities & participation and environmental factors components. *BMC Medical Informatics and Decision Making, 21*(1), 367.

Davis, H., Schlundt, D., Bonnet, K., Camarata, S., Bess, F. H., & Hornsby, B. (2021, June 1). Understanding listening-related fatigue: Perspectives of adults with hearing loss. *International Journal of Audiology, 60*(6), 458–468.

De Luca, R., Bonanno, M., Rifici, C., Quartarone, A., & Calabrò, R. S. (2023, July 13). Post-traumatic olfactory dysfunction: A scoping review of assessment and rehabilitation approaches. *Frontiers in Neurology, 14*, 1193406.

DeBonis, D., & Donohue, C. (2024, June 1). *Survey of audiology: Fundamentals for audiologists and health professionals*. CRC Press.

Diaconu, M. (2021, December 8). Being and making the olfactory self. Lessons from contemporary artistic practices. In *Olfaction: An interdisciplinary perspective from philosophy to life sciences* (pp. 55–73). Springer.

Dinarvand, D., Panthakey, J., Hassan, A., & Ahmed, M. H. (2024, November 1). Frailty and visual impairment in elderly individuals: Improving outcomes and modulating cognitive decline through collaborative care between geriatricians and ophthalmologists. *Diseases, 12*(11), 273.

D'Mello, R., & Dickenson, A. H. (2008, July 1). Spinal cord mechanisms of pain. *British Journal of Anaesthesia, 101*(1), 8–16.

Fabio, R. A., Orsino, C., Lecciso, F., Levante, A., & Suriano, R. (2024, March 1). Atypical sensory processing in adolescents with attention deficit hyperactivity disorder: A comparative study. *Research in Developmental Disabilities, 146*, 104674.

Fatuzzo, I., Niccolini, G. F., Zoccali, F., Cavalcanti, L., Bellizzi, M. G., Riccardi, G., de Vincentiis, M., Fiore, M., Petrella, C., Minni, A., & Barbato, C. (2023, January 20). Neurons, nose, and neurodegenerative diseases: Olfactory function and cognitive impairment. *International Journal of Molecular Sciences, 24*(3), 2117.

Fernandez, E. M., Oliveira, D. N., Silva-Neto, A. V., Davila, R. N., Lengler, L., Sartim, M. A., Farias, A. S., Ferreira, L. C., Carvalho, É. D., Wen, F. H., & Murta, F. (2025, January 3).

Physical and sensory long-term disabilities from Bothrops snakebite Envenomings in Manaus, Western Brazilian Amazon. *Toxins, 17*(1), 22.

Fjaeldstad, A. W., & Smith, B. (2022, June 8). The effects of olfactory loss and parosmia on food and cooking habits, sensory awareness, and quality of life—A possible avenue for regaining enjoyment of food. *Food, 11*(12), 1686.

Gamage, B., Do, T. T., Price, N. S., Lowery, A., & Marriott, K. (2023, October 22). What do blind and low-vision people really want from assistive smart devices? Comparison of the literature with a focus study. In *Proceedings of the 25th international ACM SIGACCESS conference on computers and accessibility* (pp. 1–21).

Gao, S., Dai, Y., & Nathan, A. (2022, February). Tactile and vision perception for intelligent humanoids. *Advanced Intelligent Systems, 4*(2), 2100074.

Ghazy, R. M., Mohammed AboElela, A., Abou-elyazid, H., & Al-Qahtani, A. S. (2025, September 25). Sensory impairment. In *The Palgrave encyclopedia of disability* (pp. 1–11). Springer Nature Switzerland.

Gorur, R., & Dey, J. (2021, January 1). Making the user friendly: The ontological politics of digital data platforms. *Critical Studies in Education, 62*(1), 67–81.

Haanes, G. G. (2023, December 31). Multidisciplinary approaches and community-based interventions: Adaptable strategies for managing sensory impairments in older adults. *Journal of Multidisciplinary Healthcare, 16*, 2701–2705.

Harris, J. E., & Tani, K. M. (2021). The disability frame. *University of Pennsylvania Law Review, 170*, 1663.

Hillson, R. (2017, March 17). Sensory disabilities in people with diabetes. In *Diabetes in old age* (pp. 137–147). Springer.

Hossain, M. S., Wazed, M. A., Asha, S., Hossen, M. A., Fime, S. N., Teeya, S. T., Jenny, L. Y., Dash, D., & Shimul, I. M. (2025, May). Flavor and well-being: A comprehensive review of food choices, nutrition, and health interactions. *Food Science & Nutrition, 13*(5), e70276.

Hummel, T., Liu, D. T., Müller, C. A., Stuck, B. A., Welge-Lüssen, A., & Hähner, A. (2023, March 3). Olfactory dysfunction: Etiology, diagnosis, and treatment. *Deutsches Ärzteblatt International., 120*(9), 146.

Jafari, A., Alaee, A., & Ghods, K. (2021, December 1). The etiologies and considerations of dysgeusia: A review of literature. *Journal of Oral Biosciences, 63*(4), 319–326.

Jaiswal, A., Paramasivam, A., Budhiraja, S., Santhakumaran, P., Gravel, C., Martin, J., Ogedengbe, T. O., James, T. G., Kennedy, B., Tang, D., & Tran, Y. (2024, September 5). The International Classification of Functioning, Disability and Health (ICF) core sets for deafblindness, part II of the systematic review: Linking data to the ICF categories. *European Journal of Physical and Rehabilitation Medicine, 60*(5), 893.

Juopperi, S., Sund, R., Rikkonen, T., Kröger, H., & Sirola, J. (2021, February 16). Cardiovascular and musculoskeletal health disorders associate with greater decreases in physical capability in older women. *BMC Musculoskeletal Disorders, 22*(1), 192.

Kameswaran, V., & Marathe, M. (2023, April 16). Advocacy as access work: How people with visual impairments gain access to digital banking in India. *Proceedings of the ACM on Human-Computer Interaction, 7*(CSCW1), 1–23.

Kapsalis, E., Jaeger, N., & Hale, J. (2024, April 2). Disabled-by-design: Effects of inaccessible urban public spaces on users of mobility assistive devices – A systematic review. *Disability and Rehabilitation: Assistive Technology, 19*(3), 604–622.

Koszalinski, R. S., & Olmos, B. (2022, October). Communication challenges in social isolation, subjective cognitive decline, and mental health status in older adults: A scoping review (2019–2021). *Perspectives in Psychiatric Care, 58*(4), 2741–2755.

Koziol, L. F., Budding, D. E., & Chidekel, D. (2011, December). Sensory integration, sensory processing, and sensory modulation disorders: Putative functional neuroanatomic underpinnings. *The Cerebellum, 10*(4), 770–792.

Kumar, S., Lata, S., & Verma, S. (2024, December 2). Psycho-social outcomes of disability-related stigma and social exclusion among people with physical disabilities: A systematic review. *Hellenic Journal of Psychology, 21*(3), 281–301.

Kushwaha, R., Vardhan, P. S., & Kushwaha, P. P. (2023, December 21). Chronic kidney disease interplay with comorbidities and carbohydrate metabolism: A review. *Life, 14*(1), 13.

Lamminmäki, S., Cormier, K., Davidson, H., Grigsby, J., & Sharma, A. (2023, November 15). Auditory cortex maturation and language development in children with hearing loss and additional disabilities. *Children, 10*(11), 1813.

Lauriello, M., Mazzotta, G., Mattei, A., Mulieri, I., Fioretti, A., Iacomino, E., & Eibenstein, A. (2024, May 13). Assessment of executive functions in children with sensorineural hearing loss and in children with specific language impairment: Preliminary reports. *Brain Sciences, 14*(5), 491.

Marin, C., Alobid, I., Fuentes, M., Lopez-Chacon, M., & Mullol, J. (2023, March). Olfactory dysfunction in mental illness. *Current Allergy and Asthma Reports, 23*(3), 153–164.

Marques, A. P., Ramke, J., Cairns, J., Butt, T., Zhang, J. H., Jones, I., Jovic, M., Nandakumar, A., Faal, H., Taylor, H., & Bastawrous, A. (2022, April 1). The economics of vision impairment and its leading causes: A systematic review. *EClinicalMedicine, 46*.

Matiashova, L., Hoogkamer, A. L., & Timper, K. (2023, December 25). The role of the olfactory system in obesity and metabolism in humans: A systematic review and meta-analysis. *Metabolites, 14*(1), 16.

McArthur, R., Williams, J., & Kneipp, S. (2025, June). Workplace accommodations for low-wage workers: A scoping review. *Work, 81*(2), 2444–2457.

McQuiddy, V. A., Ingram, M., Vines, M., Teeters, S., Ramstetter, A., & Strain-Riggs, S. R. (2024, July 1). Long-term impact of an occupational therapy intervention for children with challenges in sensory processing and integration. *The American Journal of Occupational Therapy, 78*(4), 7804185060.

Melis, M., Tomassini Barbarossa, I., & Sollai, G. (2023, July 31). The implications of taste and olfaction in nutrition and health. *Nutrients, 15*(15), 3412.

Molinari, G., Reale, M., Bonali, M., Anschuetz, L., Lucidi, D., Presutti, L., & Alicandri-Ciufelli, M. (2022, May). Taste impairment after endoscopic stapes surgery: Do anatomic variability of chorda tympani and surgical technique matter? Post-operative dysgeusia after EStS. *European Archives of Oto-Rhino-Laryngology, 279*(5), 2269–2277.

Okada, Y., Yoshimura, K., Toya, S., & Tsuchimochi, M. (2021, November 1). Pathogenesis of taste impairment and salivary dysfunction in COVID-19 patients. *Japanese Dental Science Review, 57*, 111–122.

Oleszkiewicz, A., Resler, K., Masala, C., Landis, B. N., Hummel, T., & Sorokowska, A. (2023, January 1). Alterations of gustatory sensitivity and taste liking in individuals with blindness or deafness. *Food Quality and Preference, 103*, 104712.

Park, W., Babushkin, V., Tahir, S., & Eid, M. (2021, March 26). Haptic guidance to support handwriting for children with cognitive and fine motor delays. *IEEE Transactions on Haptics, 14*(3), 626–634.

Petty, K. J. (2021, June). Beyond the senses: Perception, the environment, and vision impairment. *Journal of the Royal Anthropological Institute, 27*(2), 285–302.

Philip, V. S., Suryasree, P. K., & Santhosh, K. E. (2025, October 3). Efficacy and social validity of intervention using speech generating device and tactile symbols to develop communication and language in a child with visual impairment and additional disability. *Disability and Rehabilitation: Assistive Technology, 20*(7), 2557–2572.

Pineau, C., Williams, P. L., Brady, J., Waddington, M., & Frank, L. (2021, October 30). Exploring experiences of food insecurity, stigma, social exclusion, and shame among women in high-income countries: A narrative review. *Canadian Food Studies/La Revue canadienne des études sur l'alimentation, 8*(3).

Piotrowska, A., Kostyra, E., & Karaś, R. (2024). The role of senses in sensory integration in the context of child nutrition. *Roczniki Państwowego Zakładu Higieny, 75*(4), 1–9.

Powell, D. S., Oh, E. S., Lin, F. R., & Deal, J. A. (2021, August). Hearing impairment and cognition in an aging world. *Journal of the Association for Research in Otolaryngology, 22*(4), 387–403.

Putri, G. I., & Astiarani, Y. (2023, July 24). Exploring the tasting system and clinical significance of taste disorders: A narrative review: Clinical significance of taste disorder. *Journal of Urban Health Research, 1*(3), 22–36.

Rajput, S., Malviya, R., Bahadur, S., & Puri, D. (2023, October 1). Recent updates on the development of therapeutics for the targeted treatment of Alzheimer's disease. *Current Pharmaceutical Design, 29*(35), 2802–2813.

Rajput, S., Malviya, R., & Uniyal, P. (2024, October 1). Advancements in the diagnosis, prognosis, and treatment of retinoblastoma. *Canadian Journal of Ophthalmology, 59*(5), 281–299.

Rana, A., Malviya, R., Rajput, S., Sridhar, S. B., & Wadhwa, T. (2025). Trends in nanoparticle-based strategies for the management of neuroinflammation. *CNS and Neurological Disorders-Drug Targets*.

Rani, S., Malviya, R., Rajput, S., Sridhar, S. B., & Kaushik, D. (2025, October 1). Understanding the prospective of gene therapy for the treatment of breast cancer. *Revista de Senología y Patología Mamaria, 38*(4), 100683.

Roby-Brami, A., Jarrasse, N., & Parry, R. (2021, June 21). Impairment and compensation in dexterous upper-limb function after stroke. From the direct consequences of pyramidal tract lesions to behavioral involvement of both upper-limbs in daily activities. *Frontiers in Human Neuroscience, 15*, 662006.

Sahoo, S., Naskar, C., Singh, A., Rijal, R., Mehra, A., & Grover, S. (2021, September). Sensory deprivation and psychiatric disorders: Association, assessment and management strategies. *Indian Journal of Psychological Medicine, 44*(5), 436–444.

Schäfer, L., Schriever, V. A., & Croy, I. (2021, January). Human olfactory dysfunction: Causes and consequences. *Cell and Tissue Research, 383*(1), 569–579.

Section on Complementary and Integrative Medicine, Council on Children with Disabilities, Zimmer, M., Desch, L., Rosen, L. D., Bailey, M. L., Becker, D., Culbert, T. P., McClafferty, H., Sahler, O. J., & Vohra, S. (2012, June 1). Sensory integration therapies for children with developmental and behavioral disorders. *Pediatrics, 129*(6), 1186–1189.

Sharpe, L., Todd, J., Scott, A., Gatzounis, R., Menzies, R. E., & Meulders, A. (2022, March 1). Safety behaviours or safety precautions? The role of subtle avoidance in anxiety disorders in the context of chronic physical illness. *Clinical Psychology Review, 92*, 102126.

Shave, S., Botti, C., & Kwong, K. (2022, April 1). Congenital sensorineural hearing loss. *Pediatric Clinics, 69*(2), 221–234.

Shodieva, E. (2025, June 28). Sensorineural hearing impairment. *International Journal of Political Sciences and Economics, 1*(3), 168–172.

Sonneveld, M. H., & Schifferstein, H. N. (2008, January 1). The tactual experience of objects. In *Product experience* (pp. 41–67). Elsevier.

Ståhl, C., De Wispelaere, J., & MacEachen, E. (2023, May 22). The work disability trap: Manifestations, causes and consequences of a policy paradox. *Disability and Rehabilitation, 45*(11), 1916–1922.

Sterkenburg, P. S., Ilic, M., Flachsmeyer, M., & Sappok, T. (2022, December 19). More than a physical problem: The effects of physical and sensory impairments on the emotional development of adults with intellectual disabilities. *International Journal of Environmental Research and Public Health, 19*(24), 17080.

Tam, M., & Koppel, K. (2021, October). Sensory impairment: Natural result of aging. *Journal of Sensory Studies, 36*(5), e12693.

Tripathi, A., & Saranya, T. S. (2022). Issues and challenges of adults with hearing disability: A mixed-method study to compare the deaf and non-deaf adults on social adjustment. *International Journal of Health Sciences, III*, 5032–5039.

Trujillo Tanner, C., Yorgason, J. B., Richardson, S., Redelfs, A. H., Serrao Hill, M. M., White, A., Stagg, B., Ehrlich, J. R., & Markides, K. S. (2022, November 1). Sensory disabilities and social isolation among Hispanic older adults: Toward culturally sensitive measurement of social isolation. *The Journals of Gerontology: Series B, 77*(11), 2091–2100.

Tseng, Y. C., Gau, B. S., Liu, T. C., Hsieh, Y. S., Huang, G. S., & Lou, M. F. (2022, December 1). Association between sensory impairments and restricted social participation in older adults: A cross-sectional study. *Collegian, 29*(6), 850–859.

Udagatti, V. D., Rajendran, D. K., Sunny, E., Shreya, S. G., Bandela, R. K., Singh, A. K., Prabhu, S., & Kumar, P. (2023, November 1). Neural pathway of taste & its pathological conditions. *Journal of Evolution of Medical and Dental Sciences, 12*(11), 344–351.

van Dam, K., Gielissen, M., Bles, R., van der Poel, A., & Boon, B. (2024, May 18). The impact of assistive living technology on perceived independence of people with a physical disability in executing daily activities: A systematic literature review. *Disability and Rehabilitation: Assistive Technology, 19*(4), 1262–1271.

Van der Meulen, A., Hartendorp, M., Voorn, W., & Hermans, F. (2022, December 29). The perception of teachers on usability and accessibility of programming materials for children with visual impairments. *ACM Transactions on Computing Education, 23*(1), 1–21.

Völter, C., Thomas, J. P., Maetzler, W., Guthoff, R., Grunwald, M., & Hummel, T. (2021, July 26). Sensory dysfunction in old age. *Deutsches Ärzteblatt International, 118*(29–30), 512.

Waugh, A. (2013, January 8). Personal care, sensory impairment and unconsciousness. In *Foundations of nursing practice: Fundamentals of holistic care* (p. 363).

Willem, J. P. (2022, May 17). *Alzheimer's, aromatherapy, and the sense of smell: Essential oils to prevent cognitive loss and restore memory*. Simon & Schuster.

World Health Organization. (2007). *International classification of functioning, disability, and health: Children & Youth Version: ICF-CY*. World Health Organization.

World Health Organization. (2022, June 16). *World mental health report: Transforming mental health for all*. World Health Organization.

Xu, Y., Aung, H. L., Hesam-Shariati, N., Keay, L., Sun, X., Phu, J., Honson, V., Tully, P. J., Booth, A., Lewis, E., & Anderson, C. S. (2024a, August 1). Contrast sensitivity, visual field, color vision, motion perception, and cognitive impairment: A systematic review. *Journal of the American Medical Directors Association, 25*(8), 105098.

Xu, J., Sun, Y., Zhu, X., Pan, S., Tong, Z., & Jiang, K. (2024b, May 30). Tactile discrimination as a diagnostic indicator of cognitive decline in patients with mild cognitive impairment: A narrative review. *Heliyon, 10*(10), e31256.

Yang, W., Zhao, X., Chai, R., & Fan, J. (2023, August 30). Progress on mechanisms of age-related hearing loss. *Frontiers in Neuroscience, 17*, 1253574.

Yeo, B. S., Chan, J. H., Tan, B. K., Liu, X., Tay, L., Teo, N. W., & Charn, T. C. (2024, September 1). Olfactory impairment and frailty: A systematic review and meta-analysis. *JAMA Otolaryngology–Head & Neck Surgery, 150*(9), 772–783.

Younes, S. (2024, March 1). The impact of micronutrients on the sense of taste. *Human Nutrition & Metabolism, 35*, 200231.

Yu, H., Han, P., & Hummel, T. (2024, May 1). Multisensory adaptation strategies: Decreased food sensory importance in patients with olfactory dysfunction. *Food Quality and Preference, 114*, 105081.

Zárate, S. (2021). *Captioning and subtitling for d/deaf and hard of hearing audiences*. UCL Press.

Zhang, J., Huang, Y., Ye, F., Yang, B., Li, Z., & Hu, X. (2022, May 9). Evaluation of post-stroke impairment in fine tactile sensation by electroencephalography (EEG)-based machine learning. *Applied Sciences, 12*(9), 4796.

Zhou, F., Ma, Z., Rashwan, A. K., Khaskheli, M. B., Abdelrady, W. A., Abdelaty, N. S., Hassan Askri, S. M., Zhao, P., Chen, W., & Shamsi, I. H. (2024, May 22). Exploring the interplay of food security, safety, and psychological wellness in the COVID-19 era: Managing strategies for resilience and adaptation. *Food, 13*(11), 1610.

MIX
Papier aus verantwortungsvollen Quellen
Paper from responsible sources
FSC® C105338

If you have any concerns about our products,
you can contact us on
ProductSafety@springernature.com

In case Publisher is established outside the EU,
the EU authorized representative is:
Springer Nature Customer Service Center GmbH
Europaplatz 3, 69115 Heidelberg, Germany

Printed by Libri Plureos GmbH
in Hamburg, Germany